$49.75

CONTRACTOR'S GUIDE TO

QuickBooks Pro® 2005

Karen Mitchell
Craig Savage
Jim Erwin

FREE CD-ROM inside the back cover includes:

- *National Estimator*, a stand-alone estimating program with over 200 pages of manhour estimates
- *Job Cost Wizard*, to export your estimates to QuickBooks Pro
- *Show Me Video*, a 60-minute interactive tutorial

Craftsman Book Company
6058 Corte del Cedro / P.O. Box 6500 / Carlsbad, CA 92018

Acknowledgments

The authors wish to express their thanks to *Nancy Carriere,* without whose assistance this book would never have been written.

Looking for other construction reference manuals?
Craftsman has the books to fill your needs. **Call toll-free 1-800-829-8123** or write to Craftsman Book Company, P.O. Box 6500, Carlsbad, CA 92018 for a **FREE CATALOG** of over 100 books, including how-to manuals, annual cost books, and estimating software.
Visit our Web site: http://www.craftsman-book.com

Library of Congress Cataloging-in-Publication Data

Mitchell, Karen, 1962-
    Contractor's guide to QuickBooks Pro 2005 / Karen Mitchell, Craig Savage, Jim Erwin.
      p. cm.
    Includes index.
    ISBN 1-57218-156-7
    1. Construction industry--Accounting--Computer programs. 2. Construction industry--Finance--Computer programs. 3. Building--Estimates--Computer programs. 4. QuickBooks.   I. Title: Contractor's guide to Quick Books Pro 2005. II. Savage, Craig, 1947- III. Erwin, Jim, 1943- IV. Title.

HF5686.B7 M4934 2005
657'.869'02855369--dc21

                            2005043994

© 2005 Craftsman Book Company
Cover design by John Wincek

Contents

Introduction   **5**
    Why You Need This Book 5
    What You Can Expect from This Book 6
    QuickBooks, QuickBooks Pro,
       or Premier: Contractor Edition? 7
    Why Should You Believe Us? 9
    What's on the CD? 12
    How Do I Use This CD? 13

1 Setting Up Your QuickBooks Pro Company   **17**
    QuickBooks Pro Company Files 18
    Begin with Our Sample
       and Company Data Files 19
    Upgrade to QuickBooks Pro Version 2005 23
    Convert from Quicken to QuickBooks Pro 25
    Convert an Existing Data File to Our Setup . . . 30

2 How to Set QuickBooks Pro Preferences   **33**
    General Preferences 33
    Accounting Preferences 34
    Checking Preferences 36
    Finance Charge Preferences 38
    Jobs and Estimates Preferences 41
    Payroll and Employees Preferences 42
    Purchases and Vendors Preferences 43
    Reminders Preferences 45
    Reports and Graphs Preferences 46
    Sales and Customers Preferences 46
    Sales Tax and Send Forms Preferences 48
    Service Connection Preferences 50
    Spelling Preferences 51
    Tax: 1099 Preferences 52
    Time Tracking Preferences 54

3 Chart of Accounts   **55**
    How to Use the Sample Chart of Accounts . . . 57
    Change, Add to, and Print Your
       Chart of Accounts 63
    QuickBooks Premier:
       Contractor Edition Features 65
    Loan Manager 66

4 Items   **75**
    Entering Items for a Non-Inventory
       Based Business 76
    Entering Items for an Inventory
       Based Business 78
    Creating a Group of Items 80
    Entering Non-Job Related Items 81

5 Payroll Items   **85**
    Using Payroll Items to Track Workers'
       Comp Costs 85
    Using Payroll Items if You Don't Track
       Workers' Comp 93
    Including Sole Proprietor and Partners'
       Time Costs in Job Costs 96

6 Classes   **103**
    Using Classes to Track Cost Categories 103
    How to Create a Class 104

7 Customers and Jobs   **107**
    How to Set Up a Customer 107
    Adding a Job for a Customer 111
    Exporting or Printing Your Customer List . . . 114

8 Vendors and Subcontractors   **115**
    Setting Up 1099 Vendors 115
    Setting Up a Non-1099 Vendor 119

9 Employees   **121**
    Setting Up Your Employee List 121

10 Opening Balances   **129**
    Entering Opening Balances in
       QuickBooks Pro 129
    Entering Invoices for Accounts Receivable . . 130
    Entering Bills for Accounts Payable 132

11 Organizing Work Flow 135
- Setting Up Your Office Files 136
- Organizing Your Payroll 139
- Keeping Office Paperwork Current 139

12 Estimating 143
- Using a Summary Estimate You Make Outside of QuickBooks Pro 144
- Customizing an Estimate Form 146
- Memorizing an Estimate 147
- Estimates and Progress Billing 149
- Getting Detailed Estimates 149

13 Receivables 151
- Four Ways to Invoice a Customer 152
- Tracking Change Orders On Estimates 162
- How to Handle Retainage 164
- Recording a Payment You Receive 169
- Recording a Deposit 170
- Recording a Job Deposit 171

14 Payables 175
- Creating and Using Purchase Orders 176
- Using Purchase Orders to Track Multiple Draws and Committed Costs 177
- Entering Bills Without Purchase Orders 181
- Selecting Bills for Payment 183
- Printing Checks 184
- Vendor Workers' Comp Reports 186

15 Payroll 191
- Entering a Timesheet 191
- Processing Employee Payroll 195
- Allocating Sole Proprietor or Partner's Time to a Job 201

16 Using QuickBooks Pro on a Cash Basis 205
- How to Record a Check 206
- How to Record a Deposit 209
- Checking Your Transactions with the QuickBooks Pro Register 211

17 Reports 213
- How to Modify Reports 213
- Using Our Memorized Reports 217
- Using Jobs, Time & Mileage Reports 234

18 End of Month and End of Year Procedures 247
- Reading and Understanding Your Financial Reports 255
- End of Year Procedures 259

19 Real Estate Development 261
- New Accounts 261
- Setting Up a Development Job 262
- Using Items to Track Construction Costs as WIP 263
- Land Purchase Transactions 265
- Personal Loans 269
- Development Loans 275
- Construction Loans 279
- Recording the Sale of a Property 282

Appendix A   Estimating with QuickBooks Pro 287
- Three Good Reasons to Try QuickBooks Pro Estimating 287
- A Road Map to Your Destination 288
- Setting Preferences for Estimating 289
- Building Your Item List 291
- Creating an Estimate in QuickBooks Pro ...293
- Turning an Estimate into an Invoice 296
- Tidying Up Your Company File 299

Appendix B   Job Cost Tracking and Importing Estimates 301
- Estimates into Invoices 304
- Handling Tax 304
- Using Items for Job Cost Tracking 306
- Help Learning National Estimator 308
- Estimating with National Estimator 309
- Converting Estimates with *Job Cost Wizard* .323

Index 335

Introduction

Why You Need This Book

Contractor's Guide to QuickBooks Pro is a simple, hands-on guide for contractors, remodelers, subcontractors, and real estate developers who plan to use, or are now using, *QuickBooks Pro* accounting software. Think of this manual, like QuickBooks Pro, as a powerful new tool. Used correctly, it will get results you never thought possible — in ways you never imagined.

This book was written *for* contractors *by* contractors. That's why it's low on "accounting speak" and high on practical examples. We're not going to talk about debits and credits. QuickBooks Pro doesn't use those terms, so we won't either. Instead, we'll use words we all understand, like checks, estimates, bills, timecards, purchase orders, and deposit slips.

We'll help you set up a good, effective, highly-professional bookkeeping system as quickly and painlessly as possible. You'll know, at the click of a mouse button, who owes you money and who you owe. You'll know if there's enough cash on hand to pay bills. When you need a current profit and loss statement, you'll have one in a minute or two — either for the month, the year, or for just one job. When a lender or a bonding company needs a balance sheet, you'll get one in minutes.

Contractors have special payroll requirements. This book will show you how to get the most out of QuickBooks Pro's payroll system. No matter where you do business in the 50 states, QuickBooks Pro has a current tax table exactly right for your company. If QuickBooks Pro isn't doing your payroll now, we predict it will be before too long.

Since estimating is important to most construction contractors, we're going to cover estimating from three perspectives. In Chapter 12, you'll see how to enter the summary data from estimates you've already written into QuickBooks Pro. From here, you can do progress billing and job cost tracking. In Appendix A you'll learn how to use the estimating system built into QuickBooks Pro to create estimates. And for those who want still more estimating power, Appendix B has complete instructions for using Craftsman's estimating software, *National Estimator*, and *Job Cost Wizard*, a program that lets you turn estimates into invoices that will export to QuickBooks Pro. Both are included on the CD in the back of this book.

What You Can Expect from This Book

According to a recent national survey, more construction contractors use QuickBooks Pro and QuickBooks than all other accounting programs combined. When set up properly, QuickBooks Pro can handle the accounting for most small- to medium-sized (to $10 million a year) construction companies. But despite the sales hype, QuickBooks Pro isn't easy to set up and learn. Dozens of options and preference settings may lead you down the wrong road and may end in hours of frustration. We've spent hundreds of hours testing the options and preferences in QuickBooks Pro to help you get it right the first time. Follow our examples and you'll have an effective accounting system that provides all the information any successful business needs — and in the shortest time possible.

QuickBooks Pro doesn't replace accountants. But it does help organize and standardize your bookkeeping system. Every report your accountant needs is readily available. This makes it easier for your accountant to prepare tax returns and annual reports. The time saved should translate into lower costs for your company.

In this book, we'll explain how to:

- set preference options correctly for your company
- set up a Chart of Accounts that matches the work you do
- set up, edit, and use classes
- set up customers and jobs
- set up vendors and subcontractors
- set up employee payroll
- get your current account balances into QuickBooks Pro
- track transactions through QuickBooks Pro
- create and use estimates
- set up a simple and effective job cost system
- create and send invoices
- enter vendor bills
- write checks
- process payroll

- get payroll tax and workers' comp expenses into job cost reports
- get an owner's time into job cost reports (for a sole proprietorship)
- run workers' comp reports
- create and interpret job cost reports to keep track of your business
- prepare financial statements
- set up end-of-the-month and end-of-the-year procedures

In short, we'll show you how to get everything a construction company needs out of QuickBooks Pro. If you understand and apply the methods in this book, you should see real improvement in the effectiveness of your accounting system. And you'll gain the personal and financial rewards that come from working not just harder, but smarter.

QuickBooks, QuickBooks Pro or QuickBooks Premier: Contractor Edition?

You'll notice that the title of this manual refers to the "Pro" version of QuickBooks. We recommend that you use at least QuickBooks Pro because basic QuickBooks can't create estimates and can't track time spent on specific jobs. If time tracking and estimating are important to your construction company, invest the extra money to get QuickBooks Pro. If you have QuickBooks Premier or QuickBooks Premier: Contractor Edition, this book will work for you as well. All the features in this book are based on QuickBooks Pro, which is included in Premier and Contractor editions. When we refer to QuickBooks Pro, we are also referring to QuickBooks Premier and QuickBooks Premier: Contractor Edition.

You'll find another major advantage to QuickBooks Pro once we get into estimating. QuickBooks Pro can do progress billing — creating an invoice for each part of a job as you complete that part. That's important if you handle larger jobs that take weeks or even months. For example, you can send out a bill that covers 100 percent of the foundation work, 40 percent of the framing, and 20 percent of the plumbing on a job. That's an important advantage. No calculation is required. QuickBooks Pro does the math and keeps all the records: what you told the customer it would cost, what you've billed so far, and what is left to be billed. These progress billing statements may also become important business records to reduce arguments over what's still owed.

We believe QuickBooks Pro has the best combination of power and simplicity for small-volume builders and general contractors. It makes accounting simple for non-accountants because it works the way you do. You've been writing checks for years. With QuickBooks Pro, you fill in blanks on a check the same as always. But you do it on a computer screen.

And you add notes to check stubs about accounts and jobs to be charged — just the way you've always added notes to the stubs of paper checks. Timecards, purchase orders, and invoices work the same way.

Can QuickBooks Pro do the job for you? The thousands of construction companies now using QuickBooks Pro are good evidence that it can. If you're serious about making your construction business grow and prosper, you'll want the accounting and reporting power built into QuickBooks Pro.

There's an old saying among builders about construction accounting: "A builder who knows where he stands won't stand there very long." You need to know where you stand so you can make informed decisions quickly. If you agree, QuickBooks Pro may be perfect for your company.

For more than ten years, the three authors of this book have used construction accounting packages, some costing thousands of dollars. None are as slick, professional, and as easy to use as QuickBooks Pro. In our opinion, QuickBooks Pro offers the builder, remodeler, general contractor, and specialty subcontractor the best off-the-shelf accounting program on the market. It's affordable, reliable, and probably has all the features you'll need.

Even though QuickBooks Pro never mentions debits and credits, it handles accounting and reporting functions the same way an accountant would. It follows what professional bookkeepers and accountants refer to as Generally Accepted Accounting Principles (GAAP). You may never notice, but QuickBooks Pro uses conventional double-entry accounting. That means each time you enter a transaction, the numbers go two directions — one way as a debit and another way as a credit. Suppose you enter a bill from a supplier. Behind the scenes, QuickBooks Pro records the transaction two ways, first as an account payable and then as a charge to an expense category.

Strictly speaking, the way it comes out of the box, QuickBooks Pro isn't a true construction accounting program. But it's adaptable enough to fit the needs of most construction companies like a glove. For example, you can customize reports to get great job cost reports, just like a so-called "construction accounting package." That's why we feel so strongly that QuickBooks Pro is right for most construction companies.

Trial Version of QuickBooks Pro

If you want to try QuickBooks Pro before buying it, Intuit offers a free trial version on CD. Go to *www.usequickbooks.com/trials*. Select the version you are interested in and follow the prompts to either download the software or receive the trial version on CD. But don't install this trial version on a computer with any other version of QuickBooks Pro. The trial version of QuickBooks Pro 2005 isn't designed to run on the same computer with any other version of QuickBooks Pro. Installing a trial version of QuickBooks Pro can make changes in an earlier version of QuickBooks Pro that you've been using. You're not eager to deal with surprises in your existing QuickBooks Pro company data, and we don't recommend it.

Also, the trial version is only good for 15 uses. When those 15 uses are up, QuickBooks Pro stops working and will invite you to call Intuit (800-446-8848), the developer of QuickBooks Pro, and give them a charge card number to buy the program you've been trying.

If you're already using Quicken or QuickBooks, you're probably sold on QuickBooks already. You don't need a trial version. Buy the full version of QuickBooks Pro and install that program. You'll get a rebate from Intuit for upgrading to the current version.

Why Should You Believe Us?

All three authors have been in construction, using computers, and using QuickBooks and QuickBooks Pro for many years. We've helped hundreds of contractors set up and use QuickBooks and QuickBooks Pro. We're confident that what we've done for others we can do for you, too.

Karen Mitchell was a general building contractor and is currently co-owner of Online Accounting (www.onlineaccounting.com) which uses the Internet to help train contractors on QuickBooks Pro and Master Builder. Karen conducts seminars nationwide for contractors who use QuickBooks Pro. She is a frequent speaker at many construction trade shows such as: A/E/C SYSTEMS, JLC Live!, and NAHB's PCBC (Pacific Coast Builders Conference). Karen has written several books including: *Construction Forms & Contracts, Quicken for Contractors, Contractor's Guide to QuickBooks Pro, Architect's Guide to QuickBooks Pro, Interior Designer's Guide to QuickBooks Pro*, and *The Property Manager's Guide to QuickBooks Pro*.

Craig Savage has been a general building contractor, remodeler and custom homebuilder for over 25 years. He was an editor at *The Journal of Light Construction* magazine for many years, director of the *JLC LIVE! Training Shows*, Vice President of Marketing & Sales at www.BobVila.com, and most recently, VP of Marketing at Building Media, Inc.

In his own time Craig is a construction management computer consultant. He started *Construction Business Computing* and *Macintosh Construction Forum* newsletters, and his articles have appeared in *Architectural & Engineering Systems, Architectural Record, Fine Homebuilding Magazine, Computer Applications Newsletter, Remodeling News, NAHB Commercial Builder, NAHB Single Family Forum, Remodeler Magazine, Mac Week, Document Imaging, Imaging World*, and *A/E/C Computer Solutions*.

Craig is a regular speaker at the A/E/C Systems, NAHB, NARI, CSI, and PCBC annual conventions. He also instructs at seminars sponsored by the University of Wisconsin College of Engineering, and the University of California Santa Barbara Extension.

Other books he has co-written for Craftsman Book Company are *Construction Forms & Contracts*, and *Quicken for Contractors*. With Taunton Press he wrote *Trim Carpentry Techniques*.

Jim Erwin is a partner in several second-generation family-owned construction companies in upstate New York, that are involved in land development as well as residential and light commercial construction. He's an active member of the National Association of Home Builders and has written articles on using computers in construction for a variety of construction magazines. He's also the creator of GC/Works (published by sYnapse Software, Inc.), a full-featured software solution for the construction industry that uses Quicken or QuickBooks Pro as its basis.

What Comes Next?

But don't just take our word for it. Here's a brief summary of some of the features we'll show you to make QuickBooks Pro specific for your construction business needs.

Payables/Receivables

One of QuickBooks Pro's greatest strengths is in tracking money owed to you (Accounts Receivable) and money you owe others (Accounts Payable). For example, when you buy materials from a vendor and receive an invoice or delivery slip, you enter the information in a screen form called a Bill. You've seen lots of these, even if you've never used a computer and don't know anything about accounting.

QuickBooks Pro tracks payables and receivables so you can see at a glance what's owed and what's due. You can even see what's due on several different reports: Accounts Receivable Aging, Open Invoices, and Customer Balance Summary. Having all this information available almost instantly should help you sleep better at night. And it helps you make better decisions. For example, it's nice to know how long it's been since you received the last check when a customer asks for "just a little more work" on a project.

The Estimating Programs and Cost Data

QuickBooks Pro doesn't come with cost estimating data. And the estimating function built into QuickBooks Pro is limited, as you'll discover in Chapter 12. That's why we include the National Estimator and Job Cost Wizard programs. You can create an estimate with National Estimator and then import into QuickBooks Pro. Everything you need for estimating is on the CD inside the back cover of this book and gets installed when you select the "Complete" installation.

Intro-1
A National Estimator estimate imported into QuickBooks Pro using Job Cost Wizard.

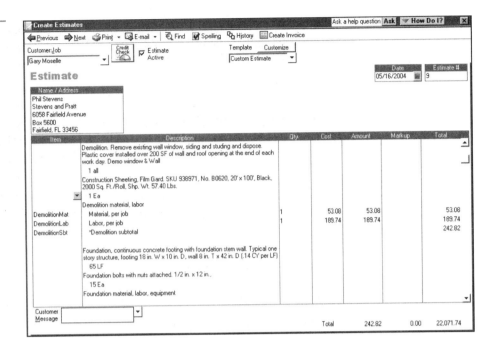

Purchase Orders

There are three good uses for the purchase orders in QuickBooks Pro. The first is to track orders placed for special order items, such as windows, doors, and skylights. Another is to track what you've committed to pay subcontractors. For example, suppose a subcontractor bids a project at a certain price. You can use a purchase order to track the bid price to compare it with the bill he submits. The third use is to establish a schedule for multiple payment draws for a subcontractor. If you agreed to pay a framing contractor, say, 30 percent on completion of the floor framing, 30 percent on completion of the walls, and 40 percent after the roof framing, you can use purchase orders to keep track of those payment schedules and payments you make against them.

Time Recording and Payroll

Payroll and time billing are flexible and sophisticated functions in QuickBooks Pro. The timecard window looks like a paper timecard, so data entry will be a familiar task. You'll find QuickBooks Pro's payroll function to be very accommodating.

QuickBooks Pro payroll handles tax withholding, additions, deductions, and company contributions. Deductions such as health insurance, payments to a retirement plan, or union dues are simple to set up and report. So are company contributions such as health, life, and workers' comp insurance paid by the company. You can record irregular payroll events such as bonuses, reimbursed travel expenses, and advances against salary. QuickBooks Pro even keeps track of sick and vacation time due employees.

QuickBooks Pro payroll reports provide all the information you need to file state and federal employer tax forms in any of the 50 states. That's part of the QuickBooks Pro payroll system. Intuit, the developers of QuickBooks Pro, have a small additional charge for the current tax table. Payroll tax tables are updated automatically as tax rates change.

If you want, you can use the QuickBooks Pro Online Payroll Service to pay all of your state and federal taxes and file all the necessary forms (including W2s) for a modest cost. You can have QuickBooks Pro print payroll checks, or use the QuickBooks Pro Online Direct Deposit Service to deposit employee paychecks automatically at nearly any bank.

What's on the CD?

We hope you bought this book because there's a CD in the back. We've loaded the CD with everything you need to get the most out of QuickBooks Pro:

- A QuickBooks Pro data file with a Chart of Accounts, items list, class list, and memorized reports for you to adapt to your business. Just plug in your own company data — vendors, subs, customers, etc. — and you're up and running.

- National Estimator, an easy-to-use estimating program with over 200 pages of construction cost estimating data for general contractors. The subtotals you create in National Estimator become cost categories in QuickBooks Pro.

- Job Cost Wizard, software that converts your National Estimator estimates into QuickBooks Pro format so you can create invoices and track costs on every job. (Requires Windows 95 or later.)

- *Show Me*, a 60-minute interactive video that shows how to use National Estimator, Job Cost Wizard, and QuickBooks Pro to estimate and track job costs.

- Sample business forms in the Forms folder. You'll find a grid to track change orders and a timecard to collect payroll data from tradesmen in the field. Other forms will help you organize information gathering in your office. Estimators may like using the estimate summary form to collect estimating data for transfer to QuickBooks Pro. To use these forms, you'll need a word processing or spreadsheet program such as Microsoft Excel, Microsoft Word, Microsoft Works, WordPerfect, or Lotus 1-2-3.

However, the CD in the back of this book doesn't include QuickBooks Pro. You won't get much out of this book without a working copy of QuickBooks Pro. So the first step will be buying and installing QuickBooks Pro if you don't have it already.

Fortunately, you can get QuickBooks Pro at most large software outlets. For the lowest price, try searching for a dealer on the Web. Some dealers offer expedited delivery at little or no cost. The CNET site lists dozens of merchants who offer QuickBooks Pro. To see the current price and terms offered by each of these merchants, go to *http://shopper.cnet.com/*.

How Do I Use This CD?

To use the CD in the back of this book, you'll need a computer running Windows 95, 98, or higher, a CD-ROM drive, and 21 to 40 Mb free on your hard drive (depending on your computer's configuration).

If you're using a version of QuickBooks Pro older than the 2005 version, many of the illustrations in this book may not look exactly like what's on your screen. That's because this manual is based on QuickBooks Pro version 2005. If you're using version 2000, 2001, 2002, 2003, or 2004, the changes will be mostly cosmetic. Version 99 and older versions are different.

Installing the CD

To install everything on the CD, put the Contractor's Guide to QuickBooks Pro CD in your CD drive (such as D:). If installation doesn't start automatically after a few seconds:

- Click **Start**
- Click **Run**
- Enter **D:\SETUP**

Intro-2
Select the correct disk letter and type setup.

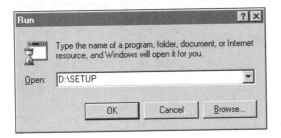

Then follow the instructions on the screen. We recommend you select the "Complete" installation when you're asked which type of installation you want to use.

When installation is complete, you'll see a new program group when you click Start and then click Programs. The new program group name is Construction Estimating. In that group, click on National Estimator 32 to start the program.

If you've installed an earlier version of National Estimator in the National folder, SETUP will automatically update obsolete files without affecting any of your estimates.

Get Help by Phone

Everything you need to know about National Estimator and Job Cost Wizard is available on the Help files that come with each program. Just click Help on the menu bar or click the question mark at the right end of the toolbar, or press the F1 key.

If you have trouble installing or using National Estimator or Job Cost Wizard, call Craftsman Book Company (Monday through Friday from 8 a.m. to 5 p.m. Pacific time) at 760-438-7828.

If you need help with QuickBooks Pro, call Intuit technical support (Monday through Friday from 6 a.m. to 4 p.m. Pacific time) at 888-320-7276 or call Online Accounting (Monday through Friday from 8 a.m. to 5 p.m. Mountain Time) at 888-254-9252.

Removing the Installed Files

To remove any of the programs installed from the Contractor's Guide to QuickBooks Pro CD from your hard drive:

- Choose **Start, Settings, Control Panel, Add/Remove Programs**.
- Click the name of the program you want to remove, **Contractor's Guide to QuickBooks Pro 2005**.
- Click **Add/Remove**.
- Click **Yes**.

Conclusion

Most contractors would agree that accounting is what they like least about running a construction business. When we started out, we felt the same way. We had a well-founded fear of accounting and an irrational loathing of computers. But using a computer for the first time isn't much different from using a Skilsaw for the first time. Treat it with respect. You'll gain confidence with every use.

We wrote this book because so many of our friends and colleagues asked us for a simple guide to setting up a construction accounting system. We've worked hard to keep it simple and still provide all the information you need. We feel the mission has been accomplished and hope you agree.

Now it's time to take the plunge. In Chapter 1, we'll wade right in by giving you some choices on how best to start using QuickBooks Pro for your company accounting.

Chapter 1
Setting Up Your QuickBooks Pro Company

Getting Started

In the introduction chapter, we mentioned that you can get a free trial version of QuickBooks Pro from the Web site *www.usequickbooks.com/trial*. This version wasn't designed to run on the same computer with any other version of QuickBooks Pro. Don't try it. But if you already have, your best choice now would be to remove the trial version. To continue, you should have installed both QuickBooks Pro 2005 and the CD in the back of this book on your computer.

We're not going to include any instructions for installing QuickBooks Pro. Intuit has done a pretty good job of that. But the disk in the back of this book is our product. So here's what you need to know about installing from the *Contractor's Guide to QuickBooks Pro* disk.

1. Insert the disk in your CD drive.

2. In a few seconds, a program should start, with Julie offering a short demonstration. Watch that demonstration, if you want. It's about six minutes long. If you're not interested, click on the red X in the lower right corner.

3. Whether you click on the red X or watch for six minutes, the demonstration ends at an installation screen.

4. Click **Install Software** and follow the instructions on the screen. We recommend the default values.

If, for any reason, Julie's demonstration doesn't start, click **Start**. Click **Run**. In the Open field, type the letter of your CD-ROM drive (such as D) followed by setup.exe. So, if D is your CD drive, you'd enter D:\setup.exe. Then click **OK**. Then just follow the on-screen installation instructions.

Once installation from the *Contractor's Guide to QuickBooks Pro* disk is complete, you'll probably want to start QuickBooks Pro. Click the **QuickBooks Pro** icon on your desktop.

QuickBooks Pro Company Files

QuickBooks Pro keeps all of your company records on disk in a single file. QuickBooks Pro refers to this file as your "company" file and that's what we'll call it in this manual. For many users, the first task is setting up a company file using QuickBooks Pro's "EasyStep Interview." We've designed a better way, as you'll soon see. And we recommend that you try it our way.

You can set up as many company data files as you want. The only requirement is that each one must have a different file name. We recommend that you use our *sample file* for practice while you experiment with QuickBooks Pro. When you've gained enough confidence to take off the training wheels, you can start your own "real" company with real records in the Company file.

Storing all company records in a single file simplifies moving your QuickBooks Pro company from one computer to another. A utility program built into QuickBooks Pro makes it easy to create a backup of any company file. We'll have more to say on making backups and moving your company file later in this book.

How to Find Your Company Data File

Where do you find your company data file? The default location is the same folder where the QuickBooks Pro program resides. But where's that? Unfortunately, there's no easy answer. The QuickBooks Pro setup program allows installation to any folder. So your company data file could be anywhere if you installed it someplace besides the default folder. Most users accept the default, however, so that's a good place to start looking. Unfortunately, there are three possible default folders.

1. If you've been using QuickBooks Pro version 5 or lower, QuickBooks Pro is probably installed on your C: drive in the qbooksw folder. If so, the new version will probably be installed in the qbooksw folder of your C: drive.

2. If you're using QuickBooks Pro version 6 or higher or you've never had any other version of QuickBooks Pro on your computer, the installation program will default to the Program Files folder, Intuit sub-folder, and QuickBooks Pro sub-sub-folder (Program Files\Intuit\QuickBooks Pro).

3. If you're using QuickBooks Premier 2004 Contractor Edition, both QBW.32.EXE and QuickBooks company and sample files are installed by default in the folder: Program Files\Intuit\QuickBooks Premier-Contractor Edition. Company files that come with this book are installed in Program Files\Intuit\QuickBooks Pro.

Company data files are kept, by default, in the same folder with the QuickBooks program (file name is QBW32.EXE). To search for your QuickBooks folder:

- Click **Start**, **Search**, then **For Files or Folders**.
- Enter QBW32.EXE, and click **Search**. Under "In Folder" you'll see the location of QBW32.EXE. Remember that folder. You may need to find it occasionally.

To find all the QuickBooks Pro company data files on your hard drive:

- Click **Start**, **Search**, then **For Files or Folders**.
- Enter *.QBW, and click **Search**. Under "In Folder" you'll see the name and location of all company files on your hard drive.

Four Choices

Where you go from here depends on your preference and what accounting program you're using now. Here are the possibilities:

- You're new to QuickBooks Pro: We suggest you use our preformatted *company.qbw* file: Start at Section 1 below — Begin With Our Sample and Company Data Files.
- You're using a prior version of QuickBooks Pro: Begin at Section 2 — Upgrading an Old QuickBooks Pro Company to Version 2005.
- You're using Quicken (another product from Intuit): Begin at Section 3 — Convert a Quicken Company Data File to QuickBooks Pro.
- You've been using QuickBooks Pro 2005 and want to set up your company data file to match our suggestions: Begin at Section 4 — Converting an Existing QuickBooks Pro 2005 Company Data File to Our Setup.

Section 1 Begin With Our Sample and Company Data Files

The people at QuickBooks have created sample company data files for a product-based company called Rock Castle Construction and a service-based company called Larry's Landscaping for demonstration purposes only. We've created a sample company data file that has all the elements of a real QuickBooks Pro construction company. We call the file *sample.qbw*. It includes:

- a Chart of Accounts for a sole proprietorship
- a list of items
- sample customers and jobs
- sample vendors
- a list of classes

- payroll items set up to track workers' compensation costs and help with the workers' comp report
- memorized transactions
- memorized reports

You can use any of these sample files for practice while you're learning how QuickBooks Pro works. Of course, we think you'll find our *sample.qbw* data file the best one to use to learn QuickBooks Pro. We've designed it specifically for the construction industry.

However, don't use any of the sample data files for your actual company data. When you're ready to begin entering actual records, you should use the company file we have on the CD called *company.qbw*. This file includes just:

- a Chart of Accounts
- items
- payroll items
- classes
- memorized reports

We recommend you modify the company data file on the CD called *company.qbw* rather than creating your own company data file from scratch.

Besides saving time, *company.qbw* is structured to prevent errors and make it easier for you to enter your own company information. To begin:

- Start QuickBooks Pro 2005.
- From the **File** menu, choose **Open Company,** or click **Open an Existing Company** in the No Company Open window.
- Select **company.qbw** (your computer may not be configured to show extensions) and then click **Open**.

If you don't see *sample.qbw* and *company.qbw* in the list of QuickBooks Pro company data files, you'll need to copy them from the CD to your QuickBooks Pro folder:

- Put the CD from inside the back cover of this book in your CD drive.
- Right click **Start** and select **Explore**.
- Scroll down the Folders side of the Windows Explorer window until you see the CD drive, probably D or E.
- Click the **+** to the left of the CD symbol to show all the folders on the CD.
- Click directly on the folder named **Examples**.

- Under Name on the right side of the Windows Explorer window you'll see the names of files in the Examples folder. Click the file you want to copy, either *sample.qbw* or *company.qbw*. The file name will be selected (white on a dark background).
- From the **Edit** menu on the Windows Explorer window, choose **Copy** to copy the file to the Windows Clipboard so you can paste it in somewhere else later.
- Now use Windows Explorer to find the QuickBooks Pro folder.
- When you've found that folder, click the folder name to select it.
- From the **Edit** menu, choose **Paste** to copy the file to your QuickBooks Pro folder.
- Click the name of the file you just pasted.
- From the **File** menu, choose **Properties**.
- Click the check mark beside **Read-only** to turn this function off so you can use the data file. Click **OK**.
- To exit the Windows Explorer window, from the **File** menu, choose **Close**.
- Start QuickBooks Pro 2005.
- From the File menu, choose **Open Company**.
- Select Company.qbw and then click **Open**.

First, let's add your company information to the new file:

- From the **Company** menu, choose **Company Information**.

Then fill in your company information:

- **Company Name** — Enter your business name.
- **Address** — Enter the address you want QuickBooks Pro to print on invoices and purchase orders.
- **Legal Name** — If you registered your business with a name different from the name in the Company box, enter it here. For example, if you're incorporated as A. C. Company but do business as Any Construction Company, enter the incorporated company name here.
- **Legal Address** — Enter the address you want QuickBooks Pro to print on legal forms.
- **First month in your fiscal year** and **First month in your income tax year** — Enter your company's first accounting month. Usually this is January.

- **Income Tax Form Used** — This depends on the type of ownership of your business. Use the drop-down list here to choose the tax form for your business's type of ownership. For example, if you own the business with someone else and it hasn't been incorporated, select Form 1065 (Partnership). If you own the business yourself and it hasn't been incorporated, select Form 1040 (Sole Proprietor). If the business has been incorporated as a regular C corporation, select Form 1120 (Corporation) or Form 1120S for an S corporation.
- Click **OK**.
- From the **File** menu, click **Close Company**.

It's important to note here that you've changed the company name that appears on the title bar of QuickBooks Pro. But you haven't changed the name of your QuickBooks Pro company data file. It's still *company.qbw*.

If you want to change the *company.qbw* data file name:

- Open the QuickBooks Pro folder.
- Click **company.qbw** using the right button on your mouse.
- From the pop-up menu, choose **Rename**.
- Enter the file name you prefer over the *company.qbw* data file name. That's the new file name of your company.

To use our *company.qbw* data file:

- From the **File** menu, choose **Open Company**.

You should see the window shown in Figure 1-1. Then:

- In the No Company Open window, select **Open an existing company**.
- In the Open Company window, click **company.qbw**.
- Click **Open**.

To use our *sample.qbw* data file:

- From the **File** menu, choose **Open Company**.

You should see the window shown in Figure 1-1. Then:

- In the No Company Open window, select **Open an existing company**.
- In the Open Company window, click **sample.qbw**.
- Click **Open**.

Now you're ready for Chapter 2, where we set up company preferences. You can skip the remainder of this chapter.

Chapter 1: Setting Up Your QuickBooks Pro Company

Figure 1-1
The No Company Open window.

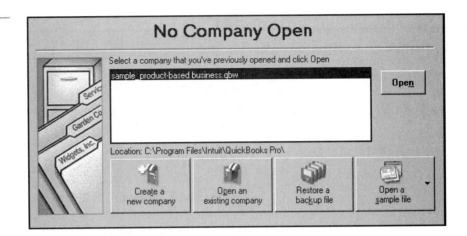

Section 2
Upgrading an Old QuickBooks Pro Company to Version 2005

To update your QuickBooks Pro company to QuickBooks Pro 2005:

- Install and start QuickBooks Pro 2005.
- In the No Company Open window, select **Open an existing company**. See Figure 1-1.
- At the Open a Company window, select your existing QuickBooks Pro file and then click **OK**.
- At the Update File to New Version window, type **YES** in the box and then click **OK**. See Figure 1-2.

Figure 1-2
Type "YES" to update your data file to QuickBooks Pro 2005.

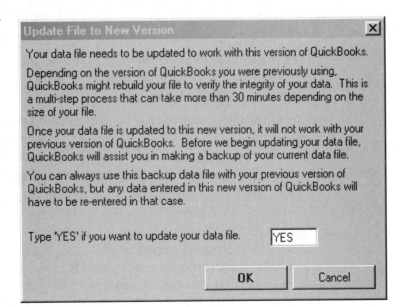

Contractor's Guide to QuickBooks Pro

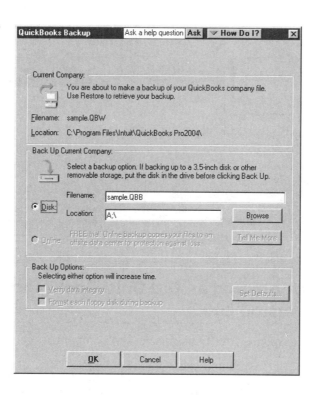

Figure 1-3
Use the Backup Company File window to specify the disk for the backup.

■ You'll get a message prompting you to back up your data. Click **OK**. If QuickBooks Pro doesn't display that message, it isn't set to open your old company file.

■ Insert your backup disk. In the Back Up Current Company: section of the window, change the location to your backup disk (usually drive A). See our example in Figure 1-3. Click **OK** to begin backing up.

QuickBooks Pro will display a message that it's updating your data. The updating program scans your data three times. The process may seem to get stuck at a certain percentage. If there's still activity on your hard drive, don't worry. Let the update continue.

When your company has been updated, you can work through the rest of the book to set up your company data. Chapters 2 through 9 will walk you through setting up QuickBooks Pro using our methods.

If your company is different from what we recommend, you may want to follow what we recommend anyway. For example, in Chapter 3 you may find that your Chart of Accounts is different from our sample Chart of Accounts. Compare your Chart of Accounts to ours to find the differences. Add new accounts where needed and consider making inactive any accounts that aren't on our recommended list. You can inactivate accounts, items, payroll items, classes, customers and jobs, vendors, and employees. Anything marked inactive isn't deleted. It just doesn't show up on the list.

Chapter 1: Setting Up Your QuickBooks Pro Company

To make an account, item, payroll item, class, customer and job, vendor, or employee inactive:

▌ From the **Lists** menu, choose the type of the particular item you want to make inactive; (ie, Chart of Accounts, Item, or Employee).

▌ Select the name of the item in the appropriate list.

▌ From the Edit menu, Select **Make Inactive**.

To see everything on the list, including anything inactive:

▌ Click **Show All** in the List window.

Now you can skip the rest of this chapter and begin by setting up your QuickBooks Pro preferences in Chapter 2.

Section 3 Convert a Quicken Company Data File to QuickBooks Pro

Quicken is a check register system. QuickBooks Pro is true double entry accounting. That's a big jump. So don't expect conversion to go flawlessly. When the automatic conversion process is complete, your company data file will still need some work. That's because Quicken and QuickBooks Pro have different "buckets" for holding information. Quicken and QuickBooks Pro are both Intuit products, but they're very different, especially in the setup procedure.

Before you begin conversion from Quicken to QuickBooks Pro, make a copy of the Quicken file for safekeeping. This is very important because you'll be making changes to the data before converting to QuickBooks Pro. Convert the copy, not your real Quicken file.

▌ Start Quicken and open your Quicken data file copy. Delete any accounts that you don't want to convert (any personal or investment accounts, for example).

▌ Put any memorized transactions you want to convert into a memorized transaction group. To find out more about how to do this, see your Quicken User's Manual.

If you've entered any customer invoices as post-dated deposits:

▌ Create and print a report of those post-dated deposits. You'll use this list later to re-enter as new invoices in your new QuickBooks Pro company data file.

▌ Delete the post-dated deposits from the Quicken checking register.

Now you're ready to close the Quicken data file and work in QuickBooks Pro:

- Close the modified Quicken file.
- Start QuickBooks Pro 2005.
- Choose **Close Company** from the **File** menu to close any open company data file.
- From the **File** menu, choose **Utilities**, and then **Convert from Quicken**.
- In the Important Documentation window, click **Convert**.
- In the Convert a Quicken File window, click the modified Quicken file, and then click **Open**. See our example in Figure 1-4.
- Enter a name for your new QuickBooks Pro file. In our example shown in Figure 1-5, we call the converted file mycompany.
- Be sure you save the file in your QuickBooks Pro directory, not the Quicken directory.
- Click **Save**.

At this point, you should carefully check the new QuickBooks Pro company data file. We've included information on the topics you should go through here. Make changes and additions as necessary.

Figure 1-4
Use this window to specify which Quicken file you want to convert to QuickBooks Pro 2005.

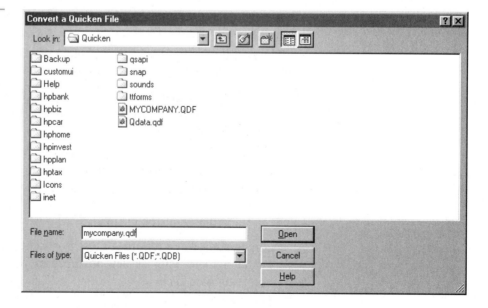

Figure 1-5
Use this window to specify the name of a new company data file.

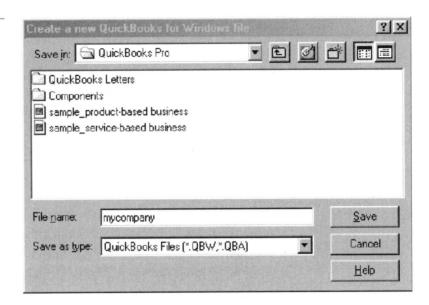

Company Information

To change company information:

▌ From the **Company** menu, choose **Company Information**.

Then fill in your company information:

▌ **Company Name** — Enter your business name.

▌ **Address** — Enter the address you want QuickBooks Pro to print on invoices and purchase orders.

▌ **Legal Name** — If you registered your business with a name different from the name in the Company box, enter it here. For example, if you're incorporated as A. C. Company but do business as Any Construction Company, enter the incorporated company name here.

▌ **Legal Address** — Enter the address you want QuickBooks Pro to print on legal forms.

▌ **First month in your fiscal year** and **First month in your income tax year** — Enter your company's first accounting month. (Usually this is January.)

▌ **Income Tax Form Used** — This depends on the type of ownership of your business. Use the drop-down list here to choose the tax form for your business's type of ownership. For example, if you own the business with someone else and it hasn't been incorporated, select Form 1065 (Partnership). If you own the business yourself and it hasn't been incorporated, select Form 1040 (Sole Proprietor). If the business has been incorporated as a C corporation, select Form 1120 (Corporation) or Form 1120S for an S corporation.

▌ Click **OK**.

Customers and Jobs

When you upgrade from Quicken to QuickBooks Pro, all of the names and addresses from Quicken usually get dumped into Customers:Jobs on the Lists menu. So you may find not only customers, but also vendors and other names in this list. If you notice this problem, first print the customer list in QuickBooks Pro:

- From the **Lists** menu, choose **Customer:Job List**.
- From the **File** menu, choose **Print List**.

To delete a name from the Customer:Job list:

- In the Customer:Job List window, click the customer name you want to delete.
- From the **Edit** menu, choose **Delete Customer:Job**. If the customer has been used in a transaction it will not allow you to delete it. Instead, you should make it Inactive.
- Click **OK**.

For more information on customers, see Chapter 7.

Vendors and Subcontractors

You can use the same process to edit your vendor and subcontractor list:

- From the **Lists** menu, choose **Vendor List**.
- From the **File** menu, choose **Print List**.

To delete a name from the Vendor list:

- In the Vendor List window, click the vendor name you want to delete.
- From the **Edit** menu, choose **Delete Vendor**.
- Click **OK**.

For more information on vendors and subcontractors, see Chapter 8.

Other Names

This is where you enter names that aren't customers, vendors, subcontractors, or employees. Use this list for contacts you deal with occasionally. To enter names on this list:

- From the **Lists** menu, choose **Other Names List**.
- From the pull-down **Other Names** menu, choose **New**.
- In the New Name window, enter information as you wish.

▌ Click **OK**.

To print the list:

▌ Choose **Print List** from the **Other Names** pull-down menu.

▌ Click **OK**.

Preferences

We'll provide complete information on setting preferences in Chapter 2.

Chart of Accounts

Chapter 3 covers setting up your Chart of Accounts. To see the QuickBooks Pro Chart of Accounts list:

▌ From the **Lists** menu, choose **Chart of Accounts**.

Items

Quicken doesn't use items, so you shouldn't have anything in your items list. For more information on setting up items, see Chapter 4. To see the QuickBooks Pro Items list:

▌ From the **Lists** menu, choose **Item List**.

Payroll Items

Since Quicken doesn't use payroll items, you shouldn't have anything in your payroll items list. For more information on setting up payroll items, see Chapter 5. To see the QuickBooks Pro Payroll Item list:

▌ From the **Lists** menu, choose **Payroll Item List**.

Classes

If you used classes for tracking jobs in Quicken, keep in mind that QuickBooks Pro's setup is different. To see the QuickBooks Pro Class list:

▌ From the **Lists** menu, choose **Class List**.

In QuickBooks Pro, customers and jobs go together. You'll need to create customers and jobs in QuickBooks Pro. Any jobs on the class list in Quicken will have to be associated with a customer in QuickBooks Pro. For more information on setting up jobs in QuickBooks Pro, see Chapter 7, Customers and Jobs. After you've moved the jobs to the correct spot, you can set up new classes. See Chapter 6.

Chapter 1: Setting Up Your QuickBooks Pro Company

Memorized Reports

If you memorized reports in Quicken, chances are those reports won't be useful in QuickBooks Pro. For more information on memorized reports, see Chapter 17, Reports.

Opening Balances

You'll want to make sure all of the balances that came over from Quicken are correct. To do this:

▌ From the **Lists** menu, choose **Chart of Accounts**.

Verify that all of the accounts are correct. If something wasn't transferred over from Quicken to QuickBooks Pro in the conversion process, you'll need to track down what's missing and enter the transaction before you enter new transactions in QuickBooks Pro. For more information on entering transactions, see Chapters 13 through 15. For more information about QuickBooks Pro opening balances, see Chapter 10, Opening Balances.

When you've finished checking and correcting your company files, go on to Chapter 2 to set up preferences.

Section 4 — Converting an Existing QuickBooks Pro 2005 Company Data File to Our Setup

To convert to our setup, you'll want to work through these chapters following our suggestions:

▌ Chapter 2 Preferences — Set up your preferences the way we suggest in Chapter 2

▌ Chapter 3 Chart of Accounts — Compare your Chart of Accounts to ours and add, change, or inactivate accounts as needed

▌ Chapter 4 Items — Add, change, or inactivate items as needed

▌ Chapter 5 Payroll Items — Add, change, or inactivate payroll items as needed

▌ Chapter 6 Classes — Add, change, or inactivate classes as needed

▌ Chapter 7 Customers — Add, change, or inactivate customers as needed

▌ Chapter 8 Vendors — Add, change, or inactivate vendors as needed

▌ Chapter 9 Employees — Add, change, or inactivate employees as needed

During this process, if you find any particular items that you want to keep but don't want to appear on any list that QuickBooks Pro makes, here's how you can inactivate them:

- Choose the list in which the item you want to inactivate appears, for example, a class in the class list.
- In the list, click the item you want to make inactive.
- From the pull-down list menu in the window, choose **Make Inactive**.
- Click **OK**.

Now the item won't show up in any list. However, it's not deleted, so you can activate it again later if you wish.

Then you'll want to go through Chapters 10 through 16 to make sure you understand how to enter transactions correctly into QuickBooks Pro. In Chapters 17 and 18 you'll see how QuickBooks Pro will make it easier for you to get the reports you need for your business.

Chapter 2
How to Set QuickBooks Pro Preferences

Preferences are used to customize QuickBooks Pro. For instance, you can turn payroll on or off, change the accounting basis for your reports, or set an interest rate for statement finance charges. There are 18 separate windows you can use to set your preferences, with several options in each window.

In this chapter we'll cover the preferences that are specific to the needs of a construction business. Any preferences not discussed aren't critical; you can set them as you like or use the defaults. We recommend you use the preferences as they default, except as noted.

Each preference has two tabs — *My Preferences* and *Company Preferences*, as in Figure 2-1. The My Preferences tab is specific to your login name and your specific needs in QuickBooks. In a multi-user environment, you can customize QuickBooks to look and act differently for each user by customizing preferences in each My Preferences tab. The Company Preferences are universal. If you change these preferences, then the preference changes company-wide, not just for an individual user.

To set your preferences, go to the **Edit** menu, select **Preferences**, and then select, one by one, the preferences shown on the following pages.

General Preferences

QuickBooks Pro defaults to the General Preferences window when you choose the Preferences command. You'll notice that the General icon is highlighted in the scroll bar of Preference icons at the left side of the window. You'll be moving along this scroll bar to select the other Preferences windows.

In the General Preferences window you can set defaults to move easily from one part of QuickBooks Pro to another, change the way the QuickBooks Pro windows look on your computer, and change several other items. The settings are really self-explanatory so you can just go ahead and set them.

- Click the defaults as you wish.
- Click the **Accounting** icon in the scroll bar to move to the next set of Preferences.

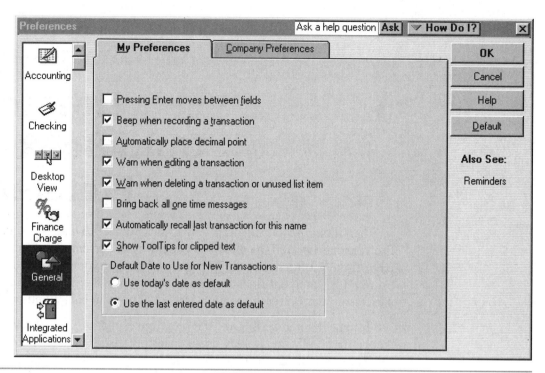

Figure 2-1
Use the My Preferences tab window to set the recommended settings for General Preferences.

If you changed the settings in the General Preferences window, QuickBooks Pro will ask if you want to save the changes. You'll be asked this question whenever you change a Preferences window and don't click **OK** to leave the window.

▌ Click **Yes** in the Save Changes window.

Accounting Preferences

Here's a list of the Accounting Preferences options and what they do:

Use account numbers — Lets you use numbers for accounts in your Chart of Accounts. In Chapter 3 we'll explain why we strongly suggest you use account numbers to identify the accounts in your Chart of Accounts. For an example of what a Chart of Accounts looks like, see our sample Chart of Accounts in Chapter 3.

Show lowest subaccount only — Displays the subaccount rather than the header accounts when you view the allocation account on a transaction. This helps the data entry operator see the proper account when viewing transactions.

Require accounts — Warns and forces you to allocate each transaction to an account.

Use class tracking — Lets you classify transactions into Labor, Materials, Subcontractors, Equipment Rental, and Other. This type of subtotaling is important when your tax preparer fills in the Cost of Goods section of your tax return. You'll probably also want to know exactly how much you've spent on these categories for each job.

Prompt to assign classes — We strongly suggest that you put a ✓ in the box. It will remind you to allocate each transaction to a class: Labor, Materials, Subcontractors, Equipment Rental, Other Job Related Costs or Overhead. See Chapter 6 — Classes for more information.

Use audit trail — Records every transaction and change made in a log file you can use if necessary. For example, you can print the audit trail report, then input the data again if your QuickBooks Pro data file gets corrupted or your hard drive fails and your backup isn't current. See Figure 2-2.

Automatically assign general journal entry number — Put a ✓ in the box. It's better to allow the system to automatically assign a number than to keep track of the number manually.

Warn when posting a transaction to Retained Earnings — We suggest this option be selected. It will protect your books from getting out of balance. The Retained Earning account balance is the total sum of money that has been made or lost since the beginning of the business. Each year QuickBooks automatically closes out your income and expenses and the difference gets posted into the Retained Earnings account. For example, let's say your first year in business ended with a net profit of $1,000. Your second year

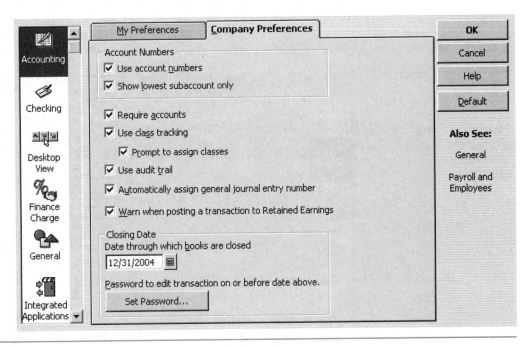

Figure 2-2
Use the Company Preferences tab window to set the recommended settings for Accounting Preferences.

in business ended with a net loss of $500. The balance in Retained Earnings should be $500. Posting a transaction to Retained Earning would cause your books to get out of balance and your tax preparer to spend hours fixing the error and costing you money.

Closing Date — Date through which books are closed — Enter the date through which you have closed the books. As soon as you submit your books to your CPA or tax preparer, you should immediately return to this screen and enter your year-end date. For example, if you've submitted paperwork or financials to your tax preparer through 12/31/04, enter that date. Then enter a password in Set Password. Entering a date and password won't prevent data entry — it'll only warn the user that the books are closed. The user will have to enter the correct password in order to post/record a transaction in a closed year.

- Click the defaults in the **Accounting** Preferences window as you wish.
- Click the **Checking** icon in the scroll bar to move to the next set of Preferences.
- Click **Yes** in the Save Changes window.

Checking Preferences

The My Preferences tab of the Checking Preferences window will help you automatically link bank accounts to specific functions or tasks. For example, let's suppose you have a payroll checking account, a general checking account, and a savings account. If you usually pay payroll from the payroll

Figure 2-3
Use the My Preferences tab of Checking Preferences to set preferences for a certain person or computer.

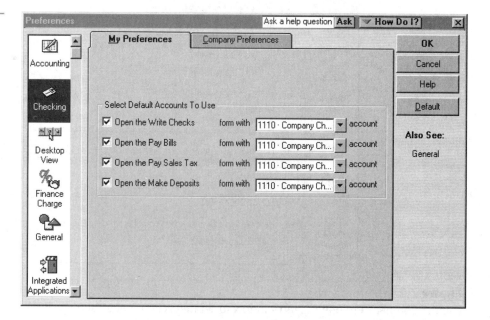

Figure 2-4
Use the Company Preferences tab of Checking Preferences to set general preferences for printing and processing checks.

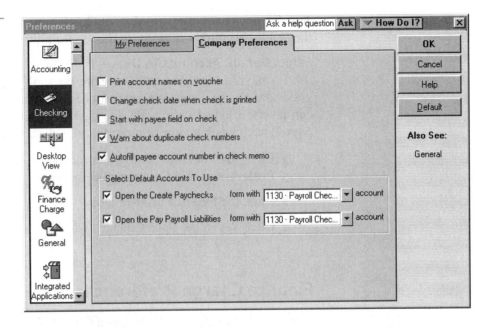

account, write checks from the general checking account, and deposit money into the savings account, you can use the My Preferences tab of the Checking Preferences window to link the appropriate bank accounts whenever you perform one of these actions. Having the correct bank account appear automatically should cut down on data entry errors. However, keep in mind that the My Preferences tab in this window lets you set preferences for a certain person or a certain computer. If you sign on with a different name, you'll need to set the preferences again for that sign-on name. You can see the setup we used in Figure 2-3.

The Company Preferences tab of the Checking Preferences window lets you set up universal preferences for printing and processing checks. See Figure 2-4.

Here's what the options shown in Figure 2-4 do:

Print account names on voucher — If you use checks with voucher attachments, the payee, date, account name, memo, amount, and total amount will appear on the voucher. If you don't select this, QuickBooks Pro omits the account name(s) but prints the rest of the information.

Change check date when check is printed — Shows the date the check was printed, not the date it was created in QuickBooks Pro. Check this if you enter your checks on a different day than you print or send them out.

Start with payee field on check — Lets you skip check number and date fields and enter information in the payee field. QuickBooks Pro will fill in the date and next check number automatically. Select this option if you're current with entering your data and stick to a consistent check number sequence.

Warn about duplicate check numbers — It's a good idea to select this to reduce the chance of duplicating a check entry.

Chapter 2: How to Set QuickBooks Pro Preferences

Autofill payee account number in check memo — If you select this box, the vendor account will print on the memo field on the check.

Select Default Accounts To Use — Enter the correct default accounts you use for payroll here. If you have more than 20 employees, we think it's a good idea to use a separate payroll checking account. We've found that it can take too long to reconcile a general checking account if you process payroll each week and print more than 20 checks for each payroll.

- Click the defaults and enter accounts as you wish in the **My Preferences** and **Company Preferences** windows.
- Click the **Finance Charge** icon.
- Click **Yes** in the Save Changes window.

Finance Charge Preferences

Now you're ready to set the Finance Charge Preferences. See Figure 2-5.

It's up to you whether or not to charge your customers a fee for late payments. Many companies use the finance charge as a threat to keep payments timely, but in practice few of them actually collect the charges. If you

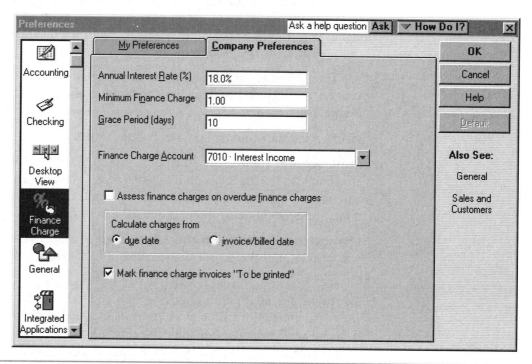

Figure 2-5
Use the Company Preferences tab window to set the recommended settings for Finance Charge Preferences.

38 Contractor's Guide to QuickBooks Pro

Figure 2-6
Use the New Account window to create an account that will track your finance charges.

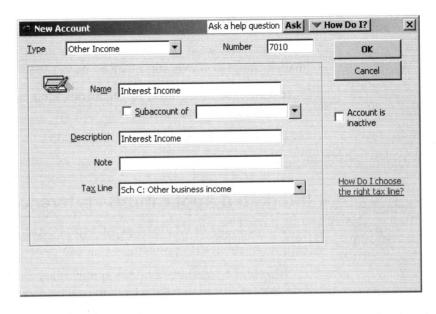

decide to charge for late payments, we suggest you explain the charges very clearly in all your contract documents so your customers aren't surprised when they're hit with these "extras."

Finance Charge Preferences let you set the percentage and minimum finance charge you want to add to overdue invoices, and the grace period. In Figure 2-5 it's set at 10 days. When you enter an amount in Minimum Finance Charge, the amount is applied regardless of the amount overdue.

Here's another example of a grace period. If your terms are net 30 and you want to start charging interest after a 30-day grace period, enter 30. Then if you send out an invoice dated January 1, QuickBooks Pro waits until January 31 to assess finance charges on any unpaid amount.

If you use finance charges, you must enter the account you use to track them. If you don't have this account in your Chart of Accounts, you need to create it now.

- At **Finance Charge Account**, pull down the menu and choose **Add New** (at the top of the list). This will bring up the New Account window. Figure 2-6 shows how we entered account 7010 Interest Income. (You don't need to add this account if you started with the Company file from the CD.)
- In **Type**, choose **Other Income** from the pull-down list.
- In **Number**, we entered **7010**. That's Interest Income from the Chart of Accounts (see Chapter 3).
- In **Name**, we entered **Interest Income**.
- In **Tax Line**, choose **Sch C: Other business income** from the pull-down list.

Chapter 2: How to Set QuickBooks Pro Preferences

■ Click **OK**.

Unless you want to play hardball, we don't suggest you select Assess finance charges on overdue finance charges. See Figure 2-5.

■ Click the **Integrated Applications** icon in the scroll bar to move to the next set of Preferences.

■ Click **Yes** in the Save Changes window.

Integrated Applications Preferences

We recommend you select QuickBooks Solutions Marketplace link to learn more about the applications that can integrate with your QBP file.

Warning! Sometimes outside applications change your QBP setup and data. We suggest you make a backup of your QuickBooks file before trying to integrate an application.

After a program has been integrated, make sure it hasn't changed your Chart of Accounts, Items, or Classes. You'll learn about your Chart of Accounts in Chapter 3, Items in Chapter 4, and Classes in Chapter 6.

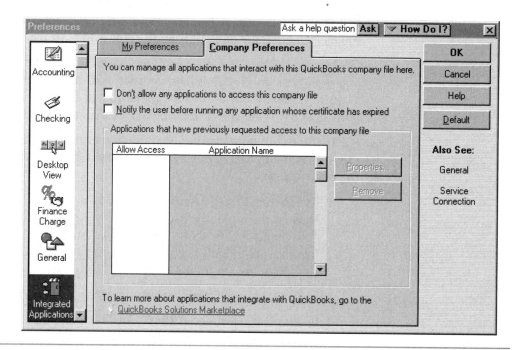

Figure 2-7
Use the Company Preferences tab window to set preferences to allow access to integrate other applications into your QBP file.

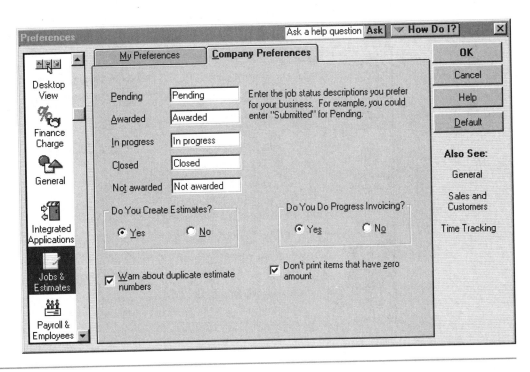

Figure 2-8
Use the Company Preferences tab window to set the recommended settings for the Jobs & Estimates Preferences.

▌ Click the **Jobs and Estimates** icon in the scroll bar to move to the next set of Preferences.

▌ Click **Yes** in the Save Changes window.

Jobs and Estimates Preferences

Whenever you choose the Customers:Jobs command on the Lists menu, you get a list of your customers, their jobs, and one of five possible descriptions for the status of each job. The default descriptions are Pending, Awarded, In progress, Closed, and Not awarded. If you want to use other descriptions for job status, you enter them on the Jobs and Estimates Preferences screen. To change a description, just enter the new description. See Figure 2-8.

▌ Set remaining defaults as you prefer. If you will be billing customers based on a percent complete by phase of construction, answer **Yes** to the question **Do You Do Progress Invoicing?** For more information on this, see Chapter 13.

▌ Click the **Payroll & Employees** Preferences icon in the scroll bar to move to the next set of Preferences.

▌ If you get the Save Changes window, click **Yes**.

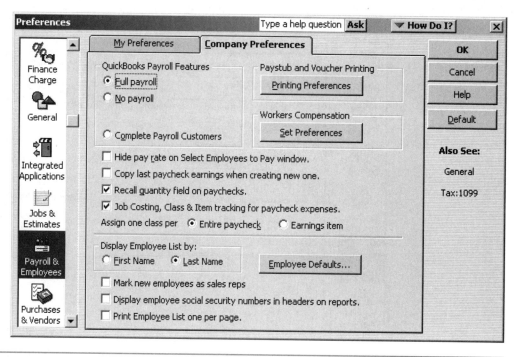

Figure 2-9
Use the Company Preferences tab window to set the recommended settings for Payroll & Employees Preferences.

Payroll and Employees Preferences

Payroll and Employee Preferences are extremely important. The way you set these preferences determines how QuickBooks Pro will associate costs, such as payroll taxes and wages, to job cost reports. See Figure 2-9 for the settings we recommend.

Full payroll features — Uses the QuickBooks Pro payroll module to allocate gross wages and payroll taxes to jobs. You'll probably want to select this even if you use a payroll service. For a more detailed discussion, see Chapter 15 on payroll.

Report all payroll taxes by Customer:Job, Service Item, and Class — Make sure you select this preference. It will automatically disburse employer payroll taxes to the job, item (job phase), and class (labor).

Assign one class per — Since we'll use one class (labor) for all transactions on a timecard, we'll select **Entire paycheck** to assign the labor class to the entire timecard. This will save you time when you enter timecards and process payroll.

You can also set the defaults you want to use when entering employee information by clicking the Employee Defaults button. Let's do that now to display the window shown in Figure 2-10.

Figure 2-10
Use the Employee Defaults window to set the defaults you'll use for each new employee you enter into QuickBooks Pro.

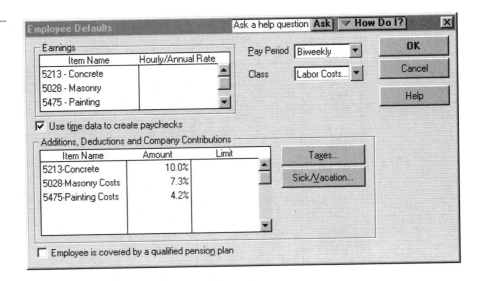

▌ Click **Employee Defaults**.

The defaults you set here will appear on the New Employee, Payroll Info Tab window each time you enter a new employee.

▌ Change the template defaults as you wish.

▌ Click **OK** to save the template.

If you are using QuickBooks Enhanced Payroll service, you will have access to a new feature that helps you prepare your workers' compensation report. Although it doesn't print the report for you, it will give you the information you'll need to fill out the report. The reason QuickBooks doesn't prepare your workers' comp report for you is because there is a vast array of different insurance companies you can purchase insurance from and each one has a different report type. To use the new feature, click **Set Preferences** in the Workers Compensation box.

Some changes you make to Payroll & Employees Preferences may cause QuickBooks Pro to advise you of the effects of the changes you've made. You can just select **OK** for these warnings.

▌ Click the **Purchases & Vendors** Preferences icon in the scroll bar to move to the next set of Preferences.

▌ Click **Yes** in the Save Changes window.

Purchases and Vendors Preferences

If your business is inventory based, you can select your preferences for inventory and purchase orders in the Purchases & Vendors Preferences window. If you want QuickBooks Pro to remind you about bills you need to

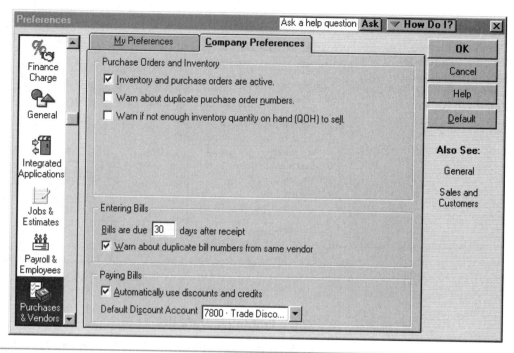

Figure 2-11
Use the Company Preferences tab window to set the recommended settings for Purchases & Vendors Preferences.

pay, you set that here. Initially QuickBooks Pro uses ten days as the lead time for paying bills and putting a message in the Reminder List. See Figure 2-11 for the Purchases and Vendors options.

Inventory and purchase orders are active — Select this if you plan to use purchase orders to track which materials are on order, when you ordered them, who you ordered them from, or if you plan to use P.O.s to track committed costs (see Chapter 14).

Warn about duplicate purchase order numbers — Select this if you want to be warned when you create a duplicate purchase order.

Bills are due — Enter a new number of days if you wish. Typically, this would be 30 days.

Warn about duplicate bill numbers — Select this if you want to be notified if you try to enter the same bill twice. This is a nice feature if you get lots of bills from certain vendors. It will help you avoid over-paying a vendor, or paying twice for the same bill.

Automatically use discounts and credits — Select this if any of your vendors let you take a discount if you pay early (or within 10 days). Enter a default account number to record all the discounts you receive. We used account number 7800 Trade Discount.

- Select items as you wish.
- Click the **Reminders** Preferences icon in the scroll bar to move to the next set of Preferences.
- Click **Yes** in the Save Changes window.

Reminders Preferences

QuickBooks Pro has an extensive set of task reminders. It can remind you to print checks, payroll checks, invoices, or to pay bills and deposit money. These can be handy functions, especially if you have several people working on your accounting system. Or they can be annoyances if you already have a schedule and live by it. QuickBooks Pro lets you "have it your way" by turning these Reminder Preferences on or off.

The Reminders Preferences screen gives you three options for each task — Show Summary, Show List, or Don't Remind Me. For some of the tasks you can set the number of days before you're warned. To follow our example, set your Reminders Preferences to match Figure 2-12 for now.

- Click the **Reports & Graphs** Preferences icon in the scroll bar to move to the next set of Preferences.
- Click **Yes** in the Save Changes window.

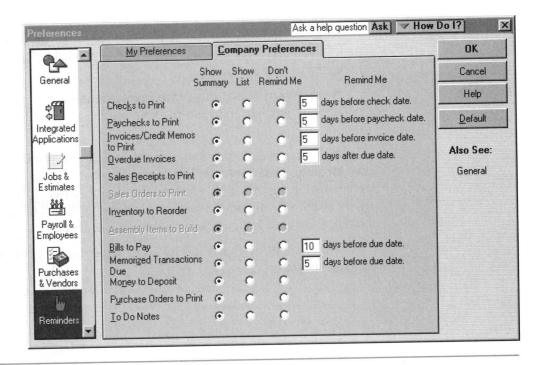

Figure 2-12
Use the Company Preferences tab window to set the recommended settings for Reminders Preferences.

Reports and Graphs Preferences

If you select Accrual in the Summary Reports Basis section of the Company Preferences tab window of Reports and Graphs Preferences, QuickBooks Pro will include unpaid vendor bills and customer invoices when figuring income and expenses, Accounts Payable, and Accounts Receivable, even though you haven't received the money. If you select Cash, only the vendor bills and customer invoices that have been paid will show up on reports.

Generally, it's a good idea to run your business on an accrual basis so you can see how much money you've earned and committed to spend over a given time period. The cash basis shows only how much money you received in a month and how much money you paid out. The accrual basis will give you a better picture of monthly income versus expenses.

For now, select the options as shown in Figures 2-13 and 2-14. You can do any fine-tuning later when you have more experience using reports.

- Click the **Sales & Customers** Preferences icon in the scroll bar to move to the next set of Preferences.

- Click **Yes** in the Save Changes window.

Sales and Customers Preferences

Many options on Sales & Customers Preferences aren't important to the construction industry, so we don't describe them here. There are three important preferences to select and we've explained below how you should make the right selection.

Select **Warn about duplicate invoice numbers** if you want to track invoices by number and want to avoid issuing two invoices with the same number.

We recommend that you don't select **Track reimbursed expenses as income**. That way you'll get reports showing customer invoices as construction income and vendor bills as job costs. If you do select it, QuickBooks Pro records income and expense on time and materials jobs by shifting amounts into, and out of, the same account.

We recommend that you DO NOT select any options in the Receive payments box.

Chapter 2: How to Set QuickBooks Pro Preferences

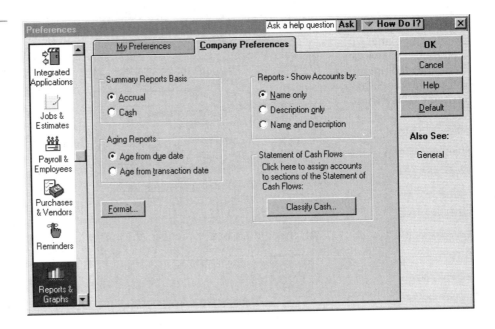

Figure 2-13
This Company Preferences tab window shows the recommended settings for Reports & Graphs.

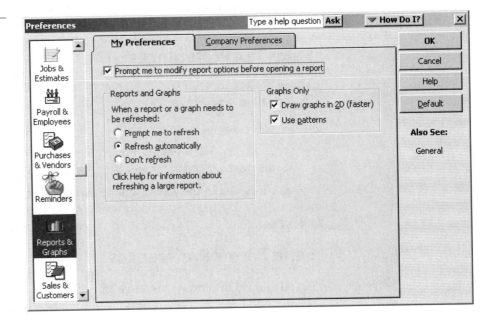

Figure 2-14
Use the My Preferences tab window to set your own preferences for Reports & Graphs.

Chapter 2: How to Set QuickBooks Pro Preferences

Figure 2-15
Use the Company Preferences tab window to set the recommended settings for the Sales & Customers Preferences.

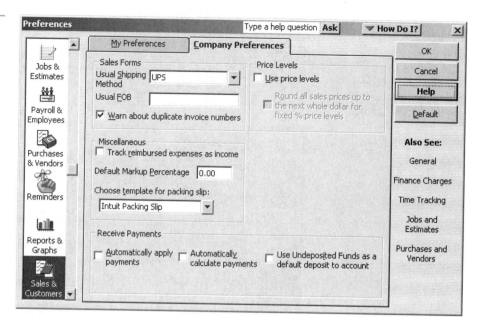

- Select items as shown in Figure 2-15.
- Click the **Sales Tax** Preferences icon in the scroll bar to move to the next set of Preferences.
- Click **Yes** in the Save Changes window.

Sales Tax Preferences

These Preferences are self-explanatory and will depend on the requirements in your state. See Figure 2-16.

- Select items as you wish.
- Click the **Send Forms** Preferences icon in the scroll bar to move to the next set of Preferences.
- Click **Yes** in the Save Changes window.

Send Forms Preferences

If you choose to send forms, such as a customer invoice, via email, you can set the Send Forms preferences to automatically attach a cover message. See Figure 2-17.

- Select items as you wish.
- Click the **Service Connection** Preferences icon in the scroll bar to move to the next set of Preferences.
- Click **Yes** in the Save Changes window.

Chapter 2: How to Set QuickBooks Pro Preferences

Figure 2-16
Use the Company Preferences tab window of Sales Tax Preferences to specify your company preferences for sales tax.

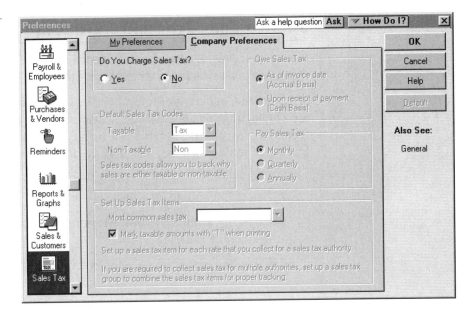

Figure 2-17
Use the Company Preferences tab window of Send Forms Preferences to set your company preferences for sending forms.

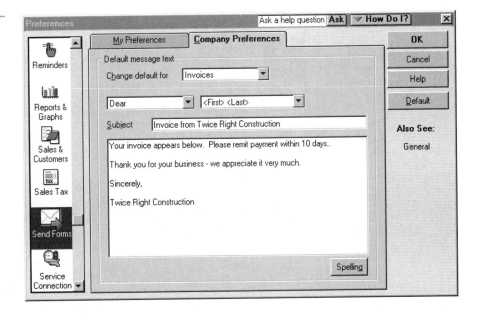

Contractor's Guide to QuickBooks Pro

Service Connection Preferences

For an overview of services provided by QuickBooks, from the **Company** menu select **Business Services Navigator**.

Service Connection Preferences lets you choose how you want to handle connections to QuickBooks services. If you're just getting started with QuickBooks, you're probably not using any of these services. In this case, leave the settings as they are and go to the next set of Preferences. See Figure 2-18.

▌ Select the items you want.

▌ Click the **Spelling Preferences** icon in the scroll bar to move to the next set of Preferences.

▌ Click **Yes** in the Save Changes window.

Figure 2-18
Service Connection preferences determine how you connect to internet applications.

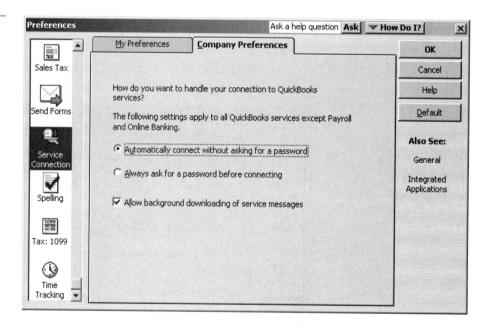

Chapter 2: How to Set QuickBooks Pro Preferences

Spelling Preferences

With the Spelling Preferences you can send invoices to customers without the worry that you've misspelled something. See Figure 2-19.

We recommend that you select **Always check spelling before printing, saving, or sending supported forms.** Then QuickBooks Pro will find any misspelled words so you can fix them before your customer finds them and begins to wonder about you.

- Select items as you wish.
- Click the **Tax:1099** Preferences icon in the scroll bar to move to the next set of Preferences.
- Click **Yes** in the Save Changes window.

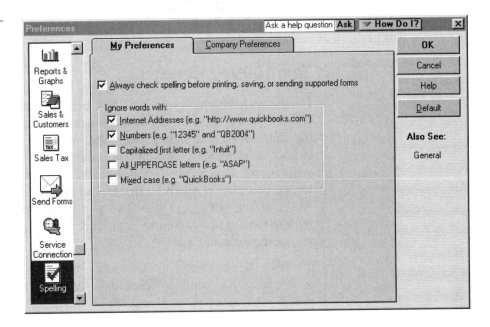

Figure 2-19
Use the My Preferences tab window of the Spelling Preferences to get QuickBooks Pro to check the spelling on your forms.

Contractor's Guide to QuickBooks Pro **51**

Figure 2-20
Use the Company Preferences tab window to specify your company preferences for 1099 tax categories.

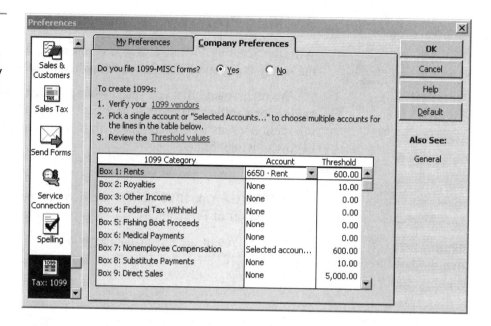

Tax: 1099 Preferences

1099s are forms that your business files with the IRS to report the total amount your company paid each vendor or subcontractor for the year starting January 1st and ending December 31st. You must file a 1099 for each vendor or subcontractor who has provided over $600 of labor during the year. If you work with vendors or subcontractors to whom you send 1099 forms, you can set up QuickBooks Pro to track all 1099-related payments. At the end of the year, QuickBooks can print your 1099 forms. You're not required to send 1099 forms to material suppliers, or to vendors from whom you purchase materials.

It's easier to complete this section if you've already set up your Chart of Accounts. So, you may want to do that first and then come back here. If you're ready now, proceed and see Figure 2-20.

▌ First answer the **Do you file 1099-MISC forms?** question.

If the answer is yes, you need to link the appropriate accounts with each 1099 category. Look in the list of 1099 categories for the categories in which you report amounts to the IRS. Most construction businesses report amounts in Box 7: Nonemployee compensation.

Chapter 2: How to Set QuickBooks Pro Preferences

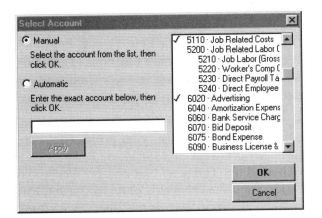

Figure 2-21
You need to select the accounts shown here, and in Figure 2-22.

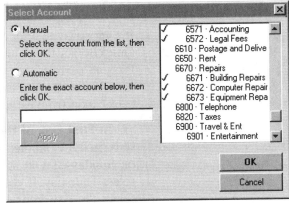

Figure 2-22
Select all accounts that you might allocate a subcontractor or outside consultant expense, so a 1099 can be created at year end.

■ Click in the **Account** column next to Box 7: Nonemployee Compensation. In the drop-down list that appears, choose the account you'll use to track 1099 payments to your vendors. If you use more than one account, click **Selected accounts** at the top of the account list and then select each individual account, as shown in Figures 2-21 and 2-22.

Note that each 1099 category must have its own unique account. For example, if you report Nonemployee Compensation (Box 7), and you have a Cost of Goods Sold account named Job Related Costs, you can select it as the account for tracking Nonemployee Compensation. But once you select Job Related Costs for Box 7, you can't use it to track any of the other 1099 categories.

■ Enter items as you wish.

■ Click the **Time Tracking** Preferences icon in the scroll bar to move to the next set of Preferences.

■ Click **Yes** in the Save Changes window.

Contractor's Guide to QuickBooks Pro 53

Chapter 2: How to Set QuickBooks Pro Preferences

Time Tracking Preferences

It's very easy to track time spent on a job using the time tracking function in QuickBooks Pro. This is one of the most important functions QuickBooks Pro can perform for you. Accurate time tracking lets you compare your actual labor to your estimated labor. In other words, it's a link in the "feedback" loop from estimate to job cost and back to estimate.

In **First Day of the Work Week** select the day your printed timecards start. In our example, we start our work week on Monday and end the week on Sunday. If you start your week on Thursday and end on Wednesday, select Thursday. If you pay your employees biweekly or on the 1st and 15th, select the day you've been using on your timecards. When QuickBooks processes payroll it will know to take the dates that make up the pay period, not the full week entered.

▪ Select items shown in Figure 2-23 as you wish.

▪ Click **OK**.

Now you should have your QuickBooks Pro company file set up and your preferences customized. The next step is to create or change your Chart of Accounts. In the next chapter we'll explain what a Chart of Accounts is and how to add, delete, or edit accounts.

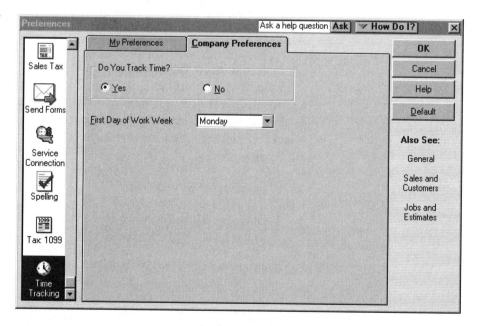

Figure 2-23
Use the Company Preferences tab window to set the recommended settings for the Time Tracking Preferences.

Chart of Accounts

The Chart of Accounts is the backbone of your accounting system. That's why it's so important that you understand how it works. You can think of the Chart of Accounts as a file cabinet with a file for each type of accounting information you want to track. For example, if you need to know how much money you spend on postage, you'll set up a file (an account in the Chart of Accounts) for Postage Expense.

The Chart of Accounts

Although you aren't required to use account numbers in your Chart of Accounts in QuickBooks Pro, we suggest that you do. Here are the standard Chart of Accounts numbers:

    1000 - 1999 Assets
    2000 - 2999 Liabilities
    3000 - 3999 Capital
    4000 - 4999 Income or Revenue
    5000 - 5999 Job Costs
    6000 - 6999 Overhead Costs
    7000 - 7999 Other Income
    8000 - 8999 Other Expense

Now let's look at each item on this list.

Assets

Assets are things your company owns. They're usually divided into two groups — current assets and fixed assets. Current assets are generally numbered from 1000 - 1499. These are assets that you can easily turn into cash, such as checking accounts, savings accounts, money market and CD accounts, Accounts Receivable, and inventory. So you might want to use account number 1100 for your company checking account because a checking account is a current asset.

Fixed assets are usually numbered from 1500 - 1999. These are items you would have to sell to generate cash. Automobiles, equipment, and land are examples of fixed assets. Let's look at an example. Last year Twice Right Construction bought a new table saw for $1,100. Since the saw cost over $500 (which is the minimum amount considered a fixed asset), the purchase was entered to an asset account rather than to an expense account.

Liabilities

Liabilities are funds your company owes. For example, last year Twice Right Construction borrowed $20,000 from We Trust U Bank. When the $20,000 loan was deposited to the checking account, the deposit was entered in the liability account Bank Loans, not an income account such as Construction Revenue.

Capital

Your capital account structure depends on whether your company is organized as a sole proprietorship, partnership, or corporation.

If your company is a *sole proprietorship*, you need a Capital account and an Owner's Drawing account. Use the Capital account to keep track of the total amount of money you've invested since starting the business, plus or minus the net profit or loss each year since you started the business. Use the Owner's Drawing account for money you take out of the business for personal use, such as checks to the grocery store, dry cleaners, ATM transactions, and your salary. It's important to keep in mind that the owner of a sole proprietorship doesn't get a regular "employee" paycheck with money deducted for payroll taxes. Instead you pay quarterly estimated taxes, which you should always allocate to the Owner's Drawing account.

If your company is a *partnership* or *LLP (Limited Liability Partnership)*, you need to set up Capital and Drawing accounts for each partner.

If your company is an *S* or *C corporation* or an *LLC*, it has a Common Stock account and sometimes a Preferred Stock account. Common stock and preferred stock represent the total sum of stock the company has issued.

Income or Revenue

Income or revenue is the income you get from the everyday way you do business, such as remodeling income or construction revenue. You would put other income or revenue, such as rents from buildings you own, in a separate Other Income account with a number between 7000 - 7999.

Job Costs

Job Costs (also called Cost of Goods Sold in the manufacturing industry) are all the costs of building your product. If you're a home builder, job costs are whatever it costs you to build a home, including the direct labor, materials, subcontractors, dump fees, and equipment rental. If you design homes, job costs include all the costs associated with designing a home, such as design labor, drafting materials, supplies, and engineering costs. If you do both designing and building, you'll have both sets of costs. It's important to note here that the *sample.qbw* and *company.qbw* data files on the CD-ROM included with this book are set up to have just *one* total Job Cost. We use classes to break down that total into labor, materials, subcontractors, equipment rental, and any other costs related to a particular job.

Overhead Costs

Overhead Costs are fixed costs you have even if you run out of work. Rent, telephone, insurance, and utilities are overhead costs.

Other Income

Other Income is income you earn outside the normal way you do business, including interest income, and gain on the sale of an asset, insurance settlement, or stock sale.

Other Expense

Other Expense is an expense that's outside of your normal business, such as a loss on the sale of an asset or stockbroker fees.

How to Use the Sample Chart of Accounts

If you haven't already done so, you may want to take a look at the Chart of Accounts in the *sample.qbw* file on the CD-ROM you got with this book. It has example transactions, a few jobs, vendors, customers, items, classes, memorized reports, memorized transactions, and a Chart of Accounts. You can use the sample company to test a new idea and get acquainted with QuickBooks Pro.

If you're ready for the real thing, you can use the Chart of Accounts in the *company.qbw* file on the CD-ROM for a jump start. This is a general Chart of Accounts for a sole proprietor building contractor. We've added other accounts we think you need for building contractor work. See the Chart of Accounts listings on the next pages for examples specific to a sole proprietor or partnership, and a corporation.

Chapter 3: Chart of Accounts

Chart of Accounts
Sole Proprietor or Partnership

Account	Type
1110 · Company Checking Account	Bank
1111 · Adjustment Register	Bank
1120 · Company Savings Account	Bank
1130 · Payroll Checking Account	Bank
1140 · Petty Cash Account	Bank
1210 · Accounts Receivable	Accounts Receivable
1300 · Inventory Asset	Other Current Asset
1310 · Employee Advances	Other Current Asset
1320 · Retentions Receivable	Other Current Asset
1330 · Security Deposit	Other Current Asset
1340 · Vendor Deposits	Other Current Asset
*1390 · Undeposited Funds	Other Current Asset
1400 · Refundable Workers Comp Deposit	Other Current Asset
1460 · Escrow Deposit	Other Current Asset
1470 · Land Purchase	Other Current Asset
1471 · Land Interest/Closing Costs	Other Current Asset
1480 · WIP - Land Development	Other Current Asset
1490 · WIP - Construction	Other Current Asset
1510 · Automobiles & Trucks	Fixed Asset
1520 · Computer & Office Equipment	Fixed Asset
1530 · Machinery & Equipment	Fixed Asset
1540 · Accumulated Depreciation	Fixed Asset
2010 · Accounts Payable	Accounts Payable
2050 · Mastercard Payable	Credit Card
2060 · Visa Card Payable	Credit Card
2100 · Payroll Liabilities	Other Current Liability
2200 · Customer Deposits	Other Current Liability
2201 · Sales Tax Payable	Other Current Liability
2240 · Worker's Comp Payable	Other Current Liability
2300 · Loans Payable	Other Current Liability
2400 · Land Aquisition Loan	Other Current Liability
2405 · Land Development Loan	Other Current Liability
2410 · Construction Loan	Long Term Liability
2460 · Truck Loan	Long Term Liability
3000 · Opening Balance Equity	Equity
**3100 · Owner's Equity	Equity
3110 · Owner's Investments	Equity
3200 · Owner's Drawing Act	Equity
3910 · Retained Earnings	Equity
3999 · Owner's Time to Jobs	Equity
4110 · Construction Income	Income
4810 · Vendor Refunds	Income
4910 · Workers' Comp Dividend	Income
5110 · Job Related Costs	Cost of Goods Sold
5200 · Job Labor Costs	Cost of Goods Sold

58   Contractor's Guide to QuickBooks Pro

Account	Type
5210 · Job Labor (Gross Wages)	Cost of Goods Sold
5220 · Worker's Compensation Costs	Cost of Goods Sold
5230 · Direct Payroll Taxes	Cost of Goods Sold
5240 · Direct Employee Benefits	Cost of Goods Sold
6020 · Advertising	Expense
6040 · Amortization Expense	Expense
6050 · Bad Debt	Expense
6060 · Bank Service Charges	Expense
6070 · Bid Deposit	Expense
6075 · Bond Expense	Expense
6090 · Business License & Fees	Expense
6100 · Car/Truck Expense	Expense
6101 · Gas & Oil	Expense
6103 · Repairs & Maintenance	Expense
6105 · Registration & License	Expense
6107 · Insurance-Auto	Expense
6130 · Cleaning/Janitorial	Expense
6135 · Computer Supplies/Equipment	Expense
6140 · Contributions	Expense
6150 · Depreciation Expense	Expense
6160 · Dues and Subscriptions	Expense
6180 · Insurance	Expense
6181 · Disability Insurance	Expense
6182 · Liability Insurance	Expense
6185 · Worker's Comp	Expense
6200 · Interest Expense	Expense
6201 · Finance Charge	Expense
6202 · Loan Interest	Expense
6203 · Credit Card Interest	Expense
6230 · Licenses and Permits	Expense
6240 · Office Expense	Expense
6490 · Office Supplies	Expense
6500 · Payroll Expenses (office)	Expense
6501 · Payroll (office staff)	Expense
6502 · Payroll Tax Expense	Expense
6508 · Vac/Holiday/Sick Pay	Expense
6509 · Employee Bonus	Expense
6510 · Employee Benefits	Expense
6570 · Professional Fees	Expense
6571 · Accounting	Expense
6572 · Legal Fees	Expense
6573 · Computer Consultants	Expense
6610 · Postage and Delivery	Expense
6650 · Rent	Expense
6670 · Repairs	Expense
6671 · Building Repairs	Expense
6672 · Computer Repairs	Expense
6673 · Equipment Repairs	Expense
6800 · Telephone	Expense
6820 · Taxes	Expense

6830 · Training and Conferences	Expense
6900 · Travel & Ent	Expense
6901 · Entertainment	Expense
6902 · Meals	Expense
6903 · Travel	Expense
6904 · Hotels/Lodging	Expense
6920 · Tools & Machinery (under $500)	Expense
6970 · Utilities	Expense
7010 · Interest Income	Other Income
7030 · Other Income	Other Income
7800 · Trade Discount	Other Income
8010 · Other Expenses	Other Expense
*2 · Purchase Orders	Non-Posting
*4 · Estimates	Non-Posting

*QuickBooks Pro sets up these accounts for you automatically. The account number for Undeposited Funds may not be the same number you listed.

**If your business is a partnership between Richard Jones and Larry Smith, this section will look like this:

3100 · Partner 1 Equity	
3110 · Richard Jones Investments	Equity
3200 · Richard Jones Drawing Act	Equity
3100 · Partner 2 Equity	
3110 · Larry Smith Investments	Equity
3200 · Larry Smith Drawing Act	Equity

Chart of Accounts
Corporation

Account	Type
1110 · Company Checking Account	Bank
1111 · Adjustment Register	Bank
1120 · Company Savings Account	Bank
1130 · Payroll Checking Account	Bank
1140 · Petty Cash Account	Bank
1210 · Accounts Receivable	Accounts Receivable
1300 · Inventory Asset	Other Current Asset
1310 · Employee Advances	Other Current Asset
1320 · Retentions Receivable	Other Current Asset
1330 · Security Deposit	Other Current Asset
1340 · Vendor Deposits	Other Current Asset
*1390 · Undeposited Funds	Other Current Asset
1400 · Refundable Workers Comp Deposit	Other Current Asset
1460 · Escrow Deposit	Other Current Asset
1470 · Land Purchase	Other Current Asset

1471 · Land Interest/Closing Costs	Other Current Asset	
1480 · WIP - Land Development	Other Current Asset	
1490 · WIP - Construction	Other Current Asset	
1510 · Automobiles & Trucks	Fixed Asset	
1520 · Computer & Office Equipment	Fixed Asset	
1530 · Machinery & Equipment	Fixed Asset	
1540 · Accumulated Depreciation	Fixed Asset	
2010 · Accounts Payable	Accounts Payable	
2050 · Mastercard Payable	Credit Card	
2060 · Visa Card Payable	Credit Card	
2100 · Payroll Liabilities	Other Current Liability	
2200 · Customer Deposits	Other Current Liability	
2201 · Sales Tax Payable	Other Current Liability	
2240 · Worker's Comp Payable	Other Current Liability	
2300 · Loans Payable	Other Current Liability	
2400 · Land Aquisition Loan	Other Current Liability	
2405 · Land Development Loan	Other Current Liability	
2410 · Construction Loan	Long Term Liability	
2460 · Truck Loan	Long Term Liability	
3000 · Opening Balance Equity	Equity	
3100 · Common Stock	Equity	
3200 · Shareholder Distribution	Equity	
3910 · Retained Earnings	Equity	
3999 · Owner's Time to Jobs	Equity	
4110 · Construction Income	Income	
4810 · Vendor Refunds	Income	
4910 · Workers' Comp Dividend	Income	
5110 · Job Related Costs	Cost of Goods Sold	
5200 · Job Labor Costs	Cost of Goods Sold	
5210 · Job Labor (Gross Wages)	Cost of Goods Sold	
5211 · Officer's Direct Labor	Cost of Goods Sold	
5220 · Worker's Compensation Costs	Cost of Goods Sold	
5230 · Direct Payroll Taxes	Cost of Goods Sold	
5240 · Direct Employee Benefits	Cost of Goods Sold	
6020 · Advertising	Expense	
6040 · Amortization Expense	Expense	
6050 · Bad Debt	Expense	
6060 · Bank Service Charges	Expense	
6070 · Bid Deposit	Expense	
6075 · Bond Expense	Expense	
6090 · Business License & Fees	Expense	
6100 · Car/Truck Expense	Expense	
6101 · Gas & Oil	Expense	
6103 · Repairs & Maintenance	Expense	
6105 · Registration & License	Expense	
6107 · Insurance-Auto	Expense	
6130 · Cleaning/Janitorial	Expense	
6135 · Computer Supplies/Equipment	Expense	
6140 · Contributions	Expense	
6150 · Depreciation Expense	Expense	

6160 · Dues and Subscriptions	Expense
6180 · Insurance	Expense
6181 · Disability Insurance	Expense
6182 · Liability Insurance	Expense
6185 · Worker's Comp	Expense
6200 · Interest Expense	Expense
6201 · Finance Charge	Expense
6202 · Loan Interest	Expense
6203 · Credit Card Interest	Expense
6230 · Licenses and Permits	Expense
6240 · Office Expense	Expense
6490 · Office Supplies	Expense
6500 · Payroll Expenses (office)	Expense
6501 · Payroll (office staff)	Expense
6502 · Payroll Tax Expense	Expense
6503 · Officer's Wages	Expense
6508 · Vac/Holiday/Sick Pay	Expense
6509 · Employee Bonus	Expense
6510 · Employee Benefits	Expense
6570 · Professional Fees	Expense
6571 · Accounting	Expense
6572 · Legal Fees	Expense
6573 · Computer Consultants	Expense
6610 · Postage and Delivery	Expense
6650 · Rent	Expense
6670 · Repairs	Expense
6671 · Building Repairs	Expense
6672 · Computer Repairs	Expense
6673 · Equipment Repairs	Expense
6800 · Telephone	Expense
6820 · Taxes	Expense
6830 · Training and Conferences	Expense
6900 · Travel & Ent	Expense
6901 · Entertainment	Expense
6902 · Meals	Expense
6903 · Travel	Expense
6904 · Hotels/Lodging	Expense
6920 · Tools & Machinery (under $500)	Expense
6970 · Utilities	Expense
7010 · Interest Income	Other Income
7030 · Other Income	Other Income
7800 · Trade Discount	Other Income
8010 · Other Expenses	Other Expense
*2 Purchase Orders	Non-Posting
*4 · Estimates	Non-Posting

*QuickBooks Pro sets up these accounts for you automatically. The account number for Undeposited Funds may not be the same number you listed.

Figure 3-1
You can change the type, number, name, description, note, tax-related information, or opening balance for any existing account in the Edit Account window.

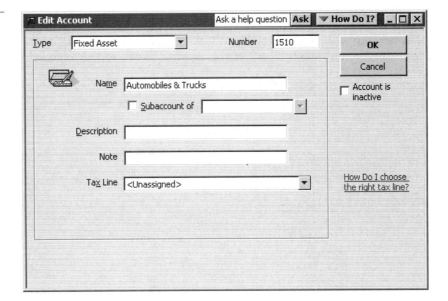

Take a little time now to compare the account list on the last 5 pages to your existing Chart of Accounts. You'll probably need to make changes for your business. For example, if you currently use expense accounts that you don't find listed in our Chart of Accounts, you'll need to add those accounts. But we'll add a word of caution here. Keep in mind that items such as payroll items and material items are linked to 5110 Job Related Costs and 4110 Construction Income accounts in the Chart of Accounts. You don't want to change these two account numbers or names.

Change, Add to, and Print Your Chart of Accounts

To change an existing account:

- From the **Lists** menu, choose **Chart of Accounts**.
- Click once to highlight the account you want to edit. Pull down the **Account** menu at the bottom of the window and choose **Edit Account**.

Figure 3-1 shows an example of the screen you'll see.

- Change the information for the account and click **OK**. For information on Opening Balances see Chapter 10.

After you change an account, you should make sure all transactions are correctly assigned. To do that:

- From the **Lists** menu, choose **Chart of Accounts** and select the account you changed.
- At the bottom of the Chart of Accounts window, pull down the **Activities** menu and choose **Use Register**.

Figure 3-2
Use the New Account window to create a new account in your Chart of Accounts.

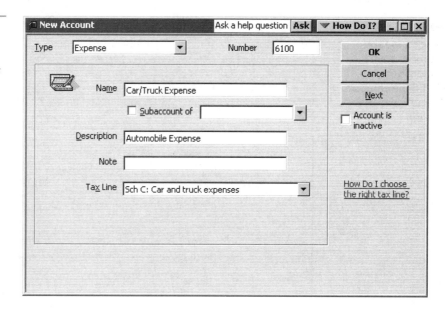

Now check the register and make sure everything listed really should be assigned to the account. If a transaction belongs in another account, double click on the transaction in the register and reassign it to the correct account.

To add a new account to your Chart of Accounts:

▌ From the **Lists** menu, choose **Chart of Accounts**.

▌ Pull down the **Account** menu at the bottom of the window and choose **New**.

▌ Add the information for the account and click **OK**.

Figure 3-2 shows an example of how to enter a new account for car/truck expenses to your Chart of Accounts.

To get more detail about a specific account, you can break it down by subaccounts. Figure 3-3 shows an example of a new subaccount window for subaccount 6101 Gas & Oil.

▌ Add the information for the subaccount and click **OK**.

After you've finished changing your Chart of Accounts, you can print a new list and keep it handy for future reference. To do that:

▌ From the pull-down **Reports** menu in the Chart of Accounts window, choose **Account Listing**.

▌ Click on **OK** to display the report.

▌ Click **Print**.

Figure 3-3
You can create subaccounts to get a more detailed breakdown of an account.

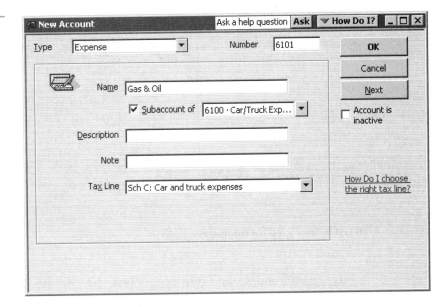

QuickBooks Premier: Contractor Edition Features

If you're using the QuickBooks Premier: Contractor (or Accountant) Edition, you have access to a few additional features, including:

- Fixed Asset Tracking
- Loan Manager
- Cash Flow Projector
- Forecasting
- Business Planning
- Expert Analysis

Fixed Asset Tracking

Fixed asset tracking allows you to record each asset you purchase in the Fixed Asset Item List located under the Lists menu. Use this list to track each item you purchase, including; property, automobiles and trucks, equipment, large tools, and computer purchases. We suggest you "write off," or expense, items that cost under $500. In other words, only use the Fixed Asset Item List for items that cost over $500. See Figure 3-4.

Unfortunately, the Fixed Asset Item List doesn't set up depreciation schedules or post depreciate for you. Instead, it gives you one convenient location to store information about an asset, such as date of purchase, purchase price, where you bought it, when, and for how much you sell the

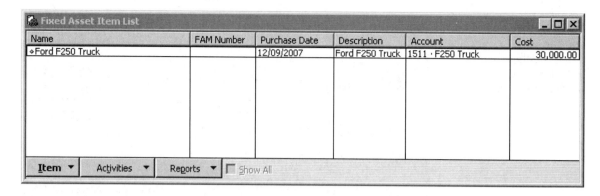

Figure 3-4
The Fixed Asset Item List is only used for items over $500.

asset, and so on. Your accountant can use the information from the fixed asset item to calculate depreciation, but you or your accountant will have to post a general journal entry to record depreciation.

Loan Manager

The Loan Manager is a payment calculator that computes the principal and interest portion of your loan payment, plus handles any fees or charges. It helps you track loans based on the information in your Long Term Liability and Other Current Liability accounts in QuickBooks. When you use the Loan Manager, you can track all of your loans in one location and be reminded of upcoming payments. Use the Loan Manager to:

- Add and remove loans you want to track.
- View payment schedules.
- Set up loan payments.
- Analyze different loan scenarios.

The Loan Manager creates payment schedules that you can view and print, allowing you to track loan-related information on a per-payment and per-total-payments basis. Plus, when you need to edit or make changes to a loan, the Loan Manager recalculates your payment information and payment schedule. Before using the Loan Manager, walk through our example below.

To set up a loan in the Loan Manager:

1. First, make sure you have a liability account set up for the loan. For example, if you took out a loan to purchase a truck, set up a loan account for the truck. See Figure 3-5.

    - From the **Lists** menu, choose **Chart of Accounts**.

Figure 3-5
Use the New Account window to create a liability account for a loan.

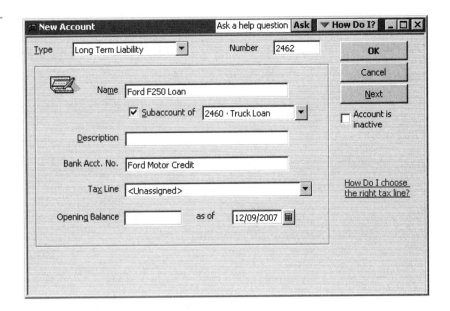

- Pull down the **Account** menu at the bottom of the window and choose **New**.
- In the Type field, select one of the following from the drop-down list:
   (a) For short-term loans (one year or less), choose Other current liability as the account type.
   (b) For long-term loans, choose Long-term liability.
- Enter the name of the lender and a description of the loan.
- Leave the opening balance at 0.00.
- Click **OK**.

2. Next, set up an asset account for the truck.
   - From the **Lists** menu, choose **Chart of Accounts**.
   - Pull down the **Account** menu at the bottom of the window and choose **New**.
   - Create the new account similar to the example in Figure 3-6.
   - Click **OK**.

3. Now we'll set up a Fixed Asset Item for the truck. Make sure you have a vendor set up for making loan payments. For example, if you'll be paying Ford Motor Credit, get that vendor entered now. For more information on entering vendors, see Chapter 8.
   - From the **Lists** menu, choose **Fixed Asset Item List**.

Chapter 3: Chart of Accounts

Figure 3-6
Use the New Account window to create a new asset account for a truck.

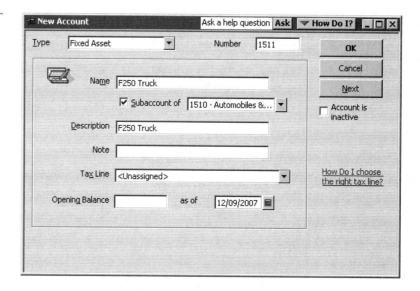

Figure 3-7
Use the New Item window to create a fixed asset item for a truck.

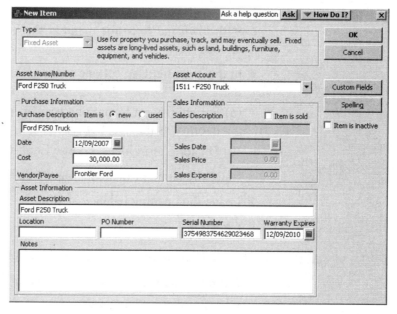

- Pull down the **Item** menu at the bottom of the window and choose **New**.
- Create the new account similar to the example in Figure 3-7.
- Click **OK**.

4. Next, we'll enter a transaction that records the purchase of the asset and the creation of a loan in the liability account you set up. For this example, let's say you purchased a truck for $30,000, put down $3,000, and took out a loan for $27,000.

- From the **Banking** menu, choose **Write Checks**.

Figure 3-8
Enter the loan amount as a negative number on the Expenses tab.

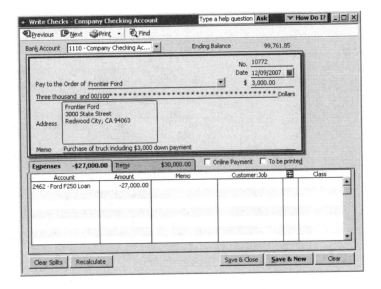

Figure 3-9
Use the Items tab of the Write Checks window to record the cost of the truck.

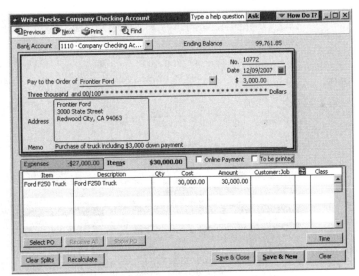

■ Fill out the top portion of the check as shown in Figure 3-8.

■ Click on the **Expenses** tab.

■ From the drop-down list in the **Account** column, select the liability account you created for the truck. In our example, it's 2462 Truck Loan. In the Amount column, enter the amount of the loan as a negative number, as shown in Figure 3-8.

■ Click on the **Items** tab.

■ From the drop-down list in the Item column, select the fixed asset item you created for the truck. In the Amount column enter the cost of the truck. See Figure 3-9.

■ Click on **Save & Close**.

Figure 3-10
Enter the account information for the loan.

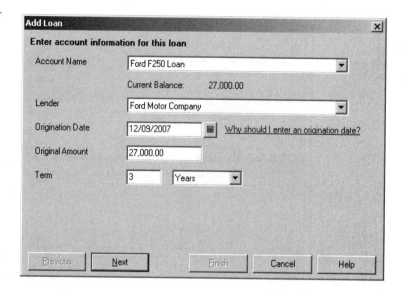

Figure 3-11
Enter the payment information for the loan.

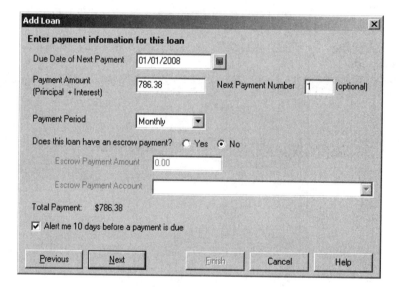

5. From the **Banking** menu, select **Loan Manager**.

6. Click on **Add a Loan**.

7. Fill in the Enter Account Information for this Loan section, as shown in Figure 3-10. Click on **Next**.

8. Fill in the Enter Payment Information for this Loan section, as in Figure 3-11. Click on **Next**.

9. Fill in the Enter Interest Information for this Loan section as shown in Figure 3-12. Click on **Next**.

10. Click **Finish**. The loan will now appear in the Loan Manager window. See Figure 3-13.

Figure 3-12
Enter the interest information for the loan.

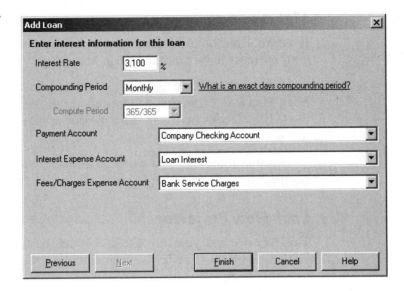

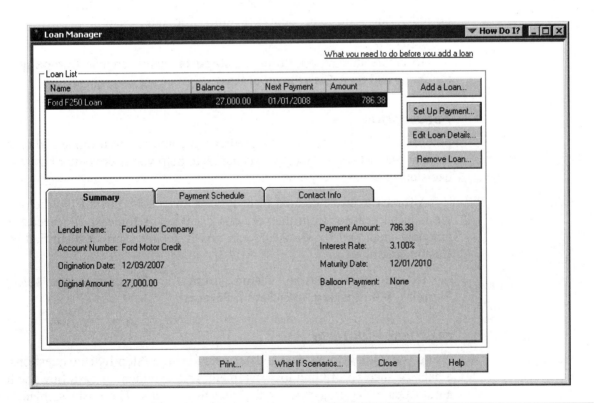

Figure 3-13
Track loan-related information in the Loan Manager.

Making Loan Payments from Loan Manager

If you're using Loan Manager to handle your loans, you should pay all of your business loans from the Loan Manager.

When you're ready to make a payment, click on **Set Up Payment** in the Loan Manager. From that window, the Loan Manager takes you directly to the Write Checks or Enter Bills windows, where you can edit your payments.

Keep in mind that you won't be allowed to automatically create a loan payment if you haven't logged in the balance of the loan as a liability.

Cash Flow Projector

This report helps you forecast how much cash you'll have by projecting your cash inflows, cash disbursements, and bank account balances on a week-by-week basis.

To see a list of the transactions that make up an amount, double-click the amount.

You can change the report date range, you can change the forecasting periods, and you can delay receipts for late customer payments. QuickBooks Help file will step you through these tasks.

To create this report, from the **Reports** menu, choose **Company & Financial** and then **Cash Flow Forecast**.

Forecasting

Forecasting allows you to make predictions about future revenue and cash flow, as well as assess "what if" scenarios to help you make better business decisions.

A forecast can be created from scratch, from actual data from the previous fiscal year, or from the previous fiscal year's forecast. A forecast is uniquely identified by its fiscal year, and if desired, further identified by Customer:Job or Class.

To use the forecasting feature, from the **Company** menu select **Planning & Budgeting**, then **Set Up Forecast**.

Business Planning

Using your QuickBooks data and answers to simple step-by-step questions, a balance sheet, profit and loss statement, and statement of cash flows will automatically be projected for the next three years. The business plan is based on the format recommended by the U.S. Small Business Administration for loan applications or a bank line of credit.

To use the Business Planning feature, from the **Company** menu, select **Planning and Budgeting**, and then **Use Business Plan Tool**.

Expert Analysis

Use the Expert Analysis tool to help you better understand your financial data and improve the performance of your business. The Expert Analysis Tool assesses performance trends for your business in important areas such as profits, sales, borrowing, liquidity, assets, and employees. It also shows how you're performing, compared to others in your industry. Comparisons are available for more than 130 specific industries.

You can find additional Help within the Expert Analysis Tool.

To use the Expert Analysis feature, from the **Company** menu, choose **Planning & Budgeting**, then choose **Use Expert Analysis Tool**.

In this chapter, we explained your Chart of Accounts. In the next chapter, Items, we'll show you how to set up items that match your job phases or divisions. Then we'll show you how to link each item to an account.

Chapter 4

Items

Although QuickBooks Pro uses the name *Items*, you can think of them as job phases — like the 16 CSI divisions, HomeTech's 28 Remodeling phases, the 31 phases in the *sample.qbw* company file, or job phases you've defined for your own business. They're like containers where you store information.

Using QuickBooks Pro Items, you can break any job into measurable, trackable units. For example, you can track your payroll and material expenses for wall framing to find out if you're over your estimated costs. It's important to do that *before* it's too late to correct any problems.

Probably no two companies will use the same set of items. The *sample.qbw* file you received with this book uses items for estimating, invoicing, and job costing. Although we've put an Items List on *company.qbw*, you really can't expect our set of items to match yours. But you can easily add, delete, and edit the items to suit your job costing needs.

One important thing to understand about items is that they link to the Chart of Accounts. When you set up a job cost item, you should link it to both a job related cost account, like 5110 Job Related Costs, and an income account, like 4110 Construction Income.

To compile a project report (Job Profitability, Job Estimates, and Job Progress, for example), you need to use items. In the *sample.qbw* data, we've created several items. You can study this list to see how we use the items in estimates and invoices shown in this book. Then you can set up items to best suit your own business. To see the Items List in *sample.qbw*:

- Open *sample.qbw*.
- From the **Lists** menu, choose **Item List**.

The project reports built into QuickBooks Pro use items to break job cost reports into smaller units — called job phases. Job phases should be the phases you actually use to estimate a job. That way you can get reports comparing your estimated costs to your actual job costs. You'll find that these reports are invaluable for running a profitable business.

Entering Items for a Non-Inventory Based Business

If you don't track inventory, you can use service inventory items to divide jobs into phases of construction. For example, if you use a specific list such as the 16 CSI divisions to make estimates, you can create that list in QuickBooks Pro using service items.

Let's see how the company on *sample.qbw*, Twice Right Construction, did it. Twice Right computerized their estimating and job costing using QuickBooks Pro items. Figure 4-1 shows some of their QuickBooks Pro items.

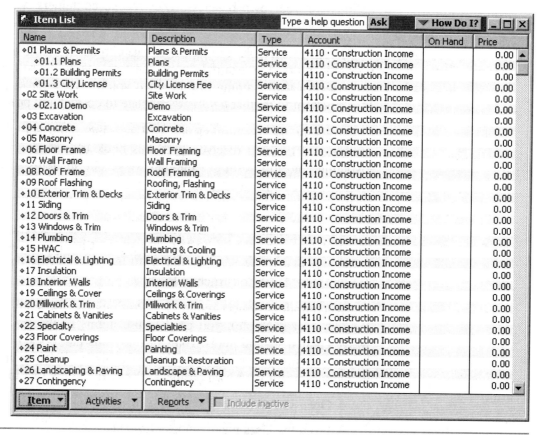

Figure 4-1
The Item List window shows information about each item created in QuickBooks Pro.

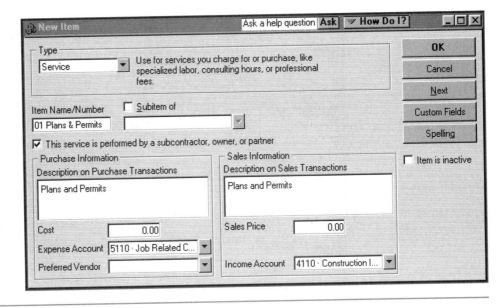

Figure 4-2
Use the New Item window to create a new service item.

To create a new service item:

- From the **Lists** menu, choose **Item List**. In the Item List window, pull down the **Item** menu and choose **New**. See our example in Figure 4-2.

- For **Type**, select **Service**. A "service" type will allow you to post labor, materials, subcontractor costs, etc. to it.

- Make sure you select the box in front of **This service is performed by a subcontractor, owner, or partner**. *This is important.* You need to select this box and the exact same accounts each time you use the item to make sure your accounting is accurate. Notice that the cost of the item is linked to the expense account you're using for Job Related Costs (5110 in our example). The price you're going to charge the customer will be linked to your Construction Income account (4110 in our example).

- Click **OK** to create the new item.

You can use subitems to do more in-depth job cost analysis. To set up a subitem, you have to create or select an item and then create its subitems. For example, using the CSI divisions to track Demo and Excavation costs as items, you could set up an item 02 Site Work. Then you can set up subitems called 02.10 Demo and 02.11 Excavation.

To create a subitem:

- From the **Lists** menu, choose **Item List**. In the Item List window, pull down the **Item** menu and choose **New**.

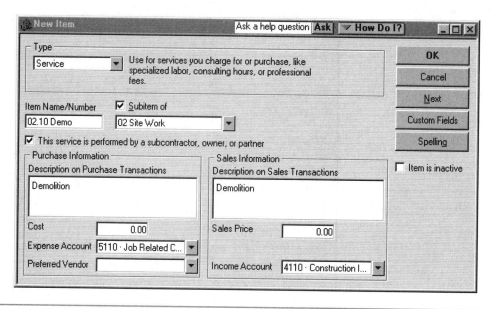

Figure 4-3
You can create subitems to get a more detailed breakdown of an item.

- For **Type**, select **Service**. A "service" type will allow you to post labor, materials, subcontractor costs, etc. to it.

- Select **This service is performed by a subcontractor, owner, or partner**. This will allow us to track estimated costs and actual costs for this item.

- Click **Subitem of** then select the new item this will be a subitem of. In our example, shown in Figure 4-3, we create the subitem 02.10 Demo of the item 02 Site Work.

- Fill in all other fields the same as you would for any item.

- Click **OK** to create the subitem.

Entering Items for an Inventory Based Business

Most general contractors and remodeling contractors don't have, or track, inventory (and in fact we discourage them from doing so). But if you're an electrical, mechanical, roofing subcontractor, or a specialty contractor you'll probably have a use for tracking inventory.

You can set up inventory items to track the materials you stock so you'll know when it's time to reorder them. You can keep track of how many parts are in stock after a sale, how many you've ordered, the cost of the goods you've sold, and the value of your inventory. You'll use purchase orders to

Chapter 4: Items

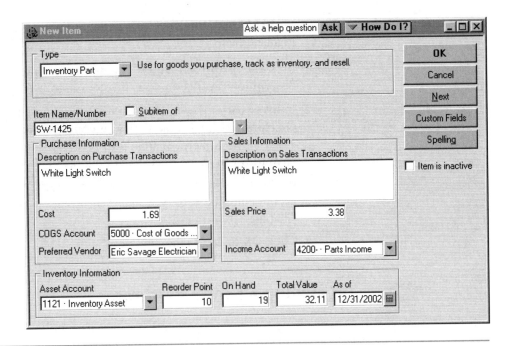

Figure 4-4
Use the New Item window to create a new inventory item.

track the materials you've bought, and invoices to record sales of materials. Just remember to use the same inventory part item on both invoices and purchase orders.

To create a new inventory item:

- From the **Lists** menu, choose **Item List**. In the Item List window, pull down the **Item** menu and choose **New**.

- For **Type**, select **Inventory Part**. See our example in Figure 4-4. If Inventory Part doesn't show up in your pull-down list under **Type**, then you'll need to change your Purchases & Vendors Preferences. To do that, from the **Edit** menu, choose **Preferences**, click **Purchases & Vendors**, click the **Company Preferences** tab, and select **Inventory and purchase orders are active**. Click **OK**.

- In the New Item window, in **Item Name/Number**, enter a short reference to uniquely identify the inventory item.

- In **Description on Purchase Transactions**, enter any description you like (but remember the description is automatically placed on a purchase order). For example, some people use the vendor's part number and description here to help the vendor identify what you're buying.

- In **Cost**, enter the cost of the item.

- In **COGS Account**, enter the appropriate Cost of Goods Sold account.

Contractor's Guide to QuickBooks Pro   79

- In **Preferred Vendor**, enter the vendor you usually order the material from.

- In **Description on Sales Transactions** enter a description. This description prefills on your invoices automatically.

- In **Inventory Information**, enter the amounts for **Reorder Point** and **On Hand**. If this is a new item in your inventory, leave **On Hand** and **Total Value** at zero. If this item is already in your inventory, you'll have to fill in these fields to get the inventory up to current levels.

- In **Asset Account** and **Income Account**, enter the appropriate account numbers. If you need information on these accounts, see Chapter 3.

- Click **OK** to create the new item.

Creating a Group of Items

You can group items together. If you use assemblies to make estimates, you can group the items in the assembly. Then you can refer to them as one unit. For example, you could make a "Plans & Permits" group item by grouping the items for plans, building permits and city license fee. Each item in the Plans & Permits group item has its own cost and selling price. When you choose an item for your estimate or invoice, you simply have to enter a quantity to get the total price for the Plans & Permits group item.

Suppose you want to use two formats for estimates: one that's an expanded detail view for your own job costing, and another that's a condensed version in a summary format that you can print out for a customer. This is especially useful if you've created a large number of subitems for each of your primary items. The condensed format might be appropriate if you're estimating the job on a fixed contract price and don't want to give the customer too much detail.

To create a new group item:

- From the **Lists** menu, choose **Item List**. In the Item List window, pull down the **Item** menu and choose **New**. See our example in Figure 4-5.

- In **Type**, select **Group**. In **Group Name/Number**, enter a name for the new group. In **Description**, enter a short description of the group you're creating.

- Select **Print items in group** if you want all the items printed when you print the group item on an invoice. This will list each item in the group individually on the invoice. If you don't select this, only the group item name will appear on the invoice.

Figure 4-5
Use grouped items to make estimating and invoicing quick and efficient.

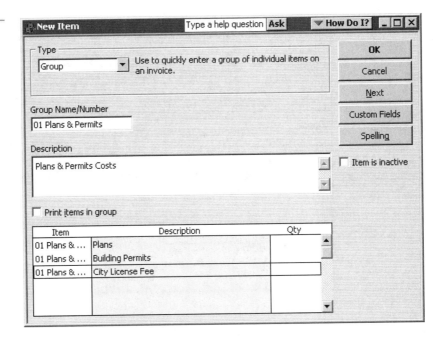

- Click in the column titled **Item**. Click the down arrow that appears and select the first item you want to include in the group. If you don't enter a quantity for the item, a value of 1 will be assumed.
- Click other items as you wish.
- Click **OK** to create the group item.

After you've created a group item, you can edit it if you need to. Any item can be in more than one group. To edit a group item:

- From the **Lists** menu, choose **Item List**.
- In the Item List window, click the group item you want to edit.
- Choose **Edit Item** from the pull-down **Item** menu and make changes in the Edit Item window.
- Click **OK**.

Entering Non-Job Related Items

In the Item List in the *sample.qbw* data file, you'll notice that near the end of the list we have several items that aren't job related. These are items that you may or may not use, such as Deposit, Beg Bal, Subtotal and Markup. You can use the Deposit item if you accept deposits from customers *and* when you subtract the deposits from the billing. For more information on customer deposits, and using subtotal items, see Chapter 13. Figure 4-6 shows how we created the Deposit item.

Chapter 4: Items

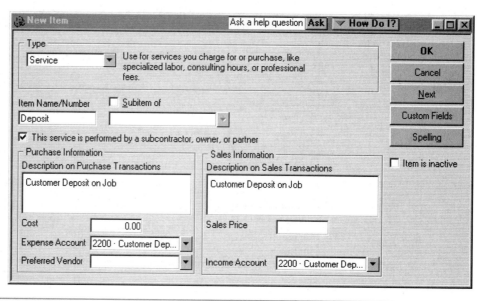

Figure 4-6
Set up the Deposit item as a template for customer deposits.

You can use the Beg Bal item when you enter your outstanding invoices — both Accounts Receivable and Accounts Payable. Figure 4-7 shows how we created the Beg Bal item.

You can use the Subtotal item if you'll be billing your customers on a time and materials basis.

▌ From **Lists** menu, select **Item List**.

▌ In the Item List Window, click on the **Item** pull-down menu and select **New**.

▌ In **Type**, select **Subtotal**, see Figure 4-8.

▌ In **Item Name/Number**, type Subtotal.

▌ Click **OK** to create the Subtotal Item.

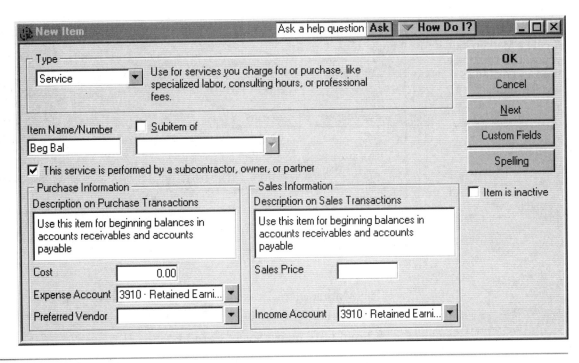

Figure 4-7
Use the Beginning Balance item to enter outstanding invoices for Accounts Receivable and Accounts Payable.

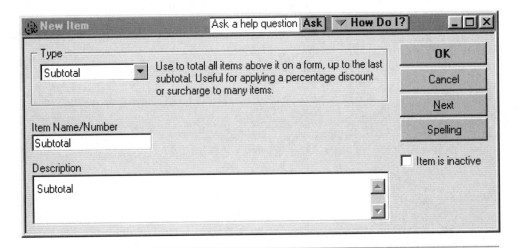

Figure 4-8
The Subtotal item is used on a time and materials invoice in Chapter 13.

Figure 4-9
Create a Markup item if you will be creating a time and materials invoice in Chapter 13.

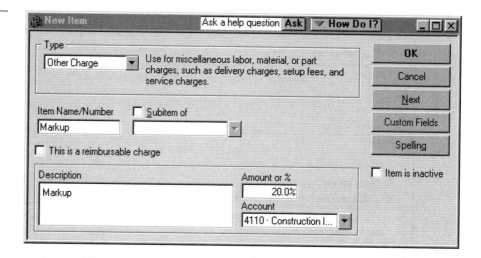

The Markup item is used if you've been billing your customers on a time and materials basis and want to add a separate line item for Markup.

- From **Lists** menu, select **Item List.**
- In the Item List Window, click on the **Item** pull-down menu and select **New.**
- In **Type**, select **Other Charge**, see Figure 4-9.
- In **Item Name/Number**, type Markup.
- Click **OK** to create the Markup Item.

After you finish setting up the items you need, you're ready to go on to Chapter 5 to set up payroll items.

Chapter 5

Payroll Items

In QuickBooks Pro, you must set up payroll items to track wages, employer and employee payroll taxes, employee and employer paid benefits, and any deductions from an employee's check. In this chapter, we'll show you how to set up the payroll items you need. In Chapter 15 (Payroll), we'll show you how to actually use the payroll items to process your payroll.

First, we'll show you how to set up payroll items to track gross wages by workers' compensation classification. This will help you prepare your workers' comp report. Second, we'll show you how to set up employer contributions to track how much you owe the insurance company for workers' comp. We'll also track the cost of the insurance by job. Third, we'll show you how to change the existing payroll items if you *don't* want to track workers' comp in QuickBooks Pro. Finally, we'll set up an hourly wage payroll item for tracking your costs to a job if you're a sole proprietor or partner of your company.

Throughout the chapter, we'll show you how to link each payroll item to the correct account in your Chart of Accounts to make sure your financial reports will be accurate. The account numbers in our examples are from the Chart of Accounts in the data file *sample.qbw* on the CD in the back of this book. If your Chart of Accounts uses different numbers, you'll need to use your own numbers.

Using Payroll Items to Track Workers' Comp Costs

Workers' comp insurance covers work-related accidents, and it's mandatory in some states. Usually, the insurance company will require you to track gross wages paid according to their classification system. They'll give you the name and rate for each classification. This insurance can be very costly for your business. For example, the roofing classification can be as high as 54 percent of wages. That means that for every $100 you pay an employee working on roofing, you'll have to pay your insurance carrier $54 for workers' comp insurance coverage. It's critical that you show these costs in your job cost reports.

To track workers' comp costs, you need to set up a payroll item for each workers' comp classification your company uses. For example, if your workers' comp provider uses a classification of 5028, you can set up an hourly rate payroll item with the same code number and call it 5028 Masonry. Then, when you enter hours for an employee who did masonry work, you choose that payroll item.

When you run a payroll report, you'll get the payroll broken down by each workers' comp classification you set up with a payroll item.

Setting Up a Payroll Item for a Workers' Comp Classification

Here's how to set up an hourly wage payroll item for a workers' comp wage classification:

- From the **Lists** menu, choose **Payroll Item List**. See Figure 5-1.

- A window may appear asking if you would like to learn more about QuickBooks Payroll Services. Select **No**.

- Click **Payroll Item** and choose **New**.

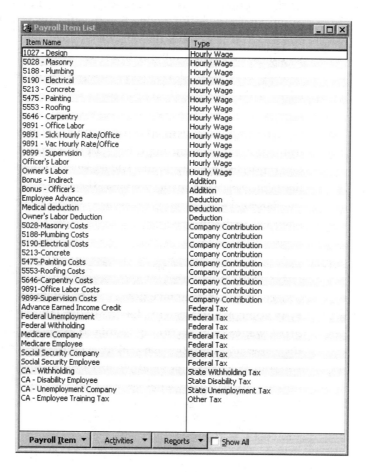

Figure 5-1
This Payroll Item List displays the names and types of various payroll items created in QuickBooks Pro.

86 *Contractor's Guide to QuickBooks Pro*

Chapter 5: Payroll Items

Figure 5-2
You need to create an hourly wage payroll item for each workers' comp classification.

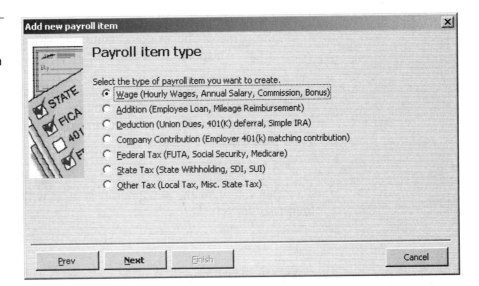

- In Select setup method, choose **Custom Setup**. Click **Next**.
- In Payroll item type window, select **Wage** (Hourly Wages, Annual Salary, Commission, Bonus), as shown in Figure 5-2.
- Click **Next**.
- In the Wages window, select **Hourly Wages**, as shown in Figure 5-3.
- Click **Next**.
- In the Wages window, select **Regular Pay**, as shown in Figure 5-4.
- Click **Next**.

Figure 5-3
In this example, we've specified hourly wage for the workers' comp payroll item we're entering.

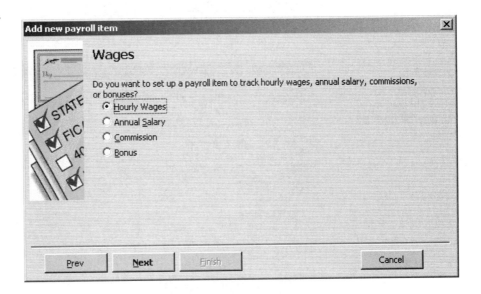

Contractor's Guide to QuickBooks Pro 87

Chapter 5: Payroll Items

- In the Name used in paychecks and payroll reports window, enter the name of the workers' comp classification. In our example, it's 5028 - Masonry. See Figure 5-5.
- Click **Next**.
- In the Expense account window, enter the account number for tracking this expense. In our example, it's 5210 - Job Labor (Gross Wages). See Figure 5-6.
- Click **Finish**.

After you enter the new payroll item, it will be added to your Payroll Item List.

Figure 5-4
Select regular pay for each workers' comp payroll item.

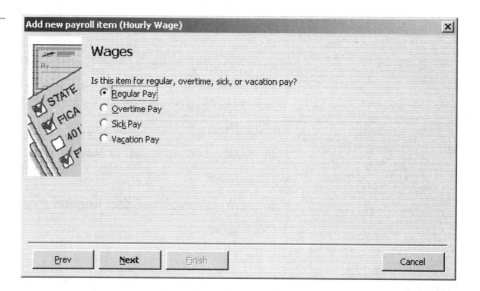

Figure 5-5
It is important to accurately classify each payroll item by name.

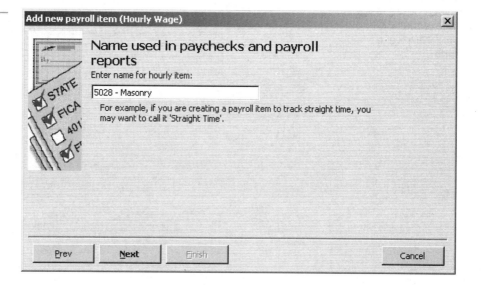

88   Contractor's Guide to QuickBooks Pro

Chapter 5: Payroll Items

Figure 5-6
In this example, 5210 - Job Labor (Gross Wages) is the correct expense account.

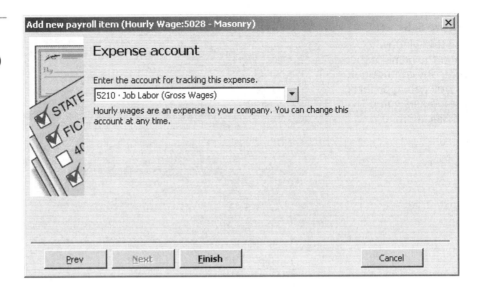

Setting Up a Payroll Item for a Company Contribution to a Workers' Comp Classification

Along with the payroll item 5028 - Masonry, you also need to set up a company contribution payroll item for 5028 - Masonry Costs. Here's how to do that:

- From the **Lists** menu, choose **Payroll Item List**.
- In the Payroll Item List window, pull down the **Payroll Item** menu and choose **New**.
- Click **Custom Setup** and **Next**.
- In the Payroll item type window, select **Company Contribution**. See Figure 5-7.

Figure 5-7
You also need to set up a Company Contribution payroll item type when you set up a workers' comp classification payroll item.

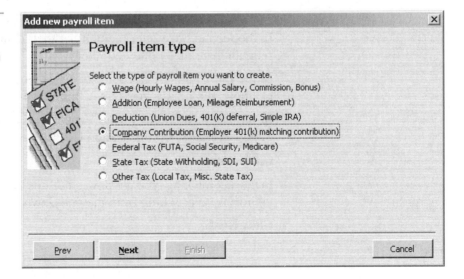

Contractor's Guide to QuickBooks Pro 89

Chapter 5: Payroll Items

Figure 5-8
Use this window to enter the name of the company contribution payroll item type for the workers' comp classification.

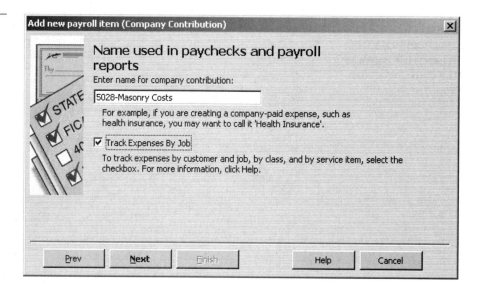

- Click **Next**.

- In the Name used in paychecks and payroll reports window, enter your workers' comp code and code name. We entered 5028 - Masonry Costs.

  Tip! The name of the company contribution has to be slightly different from the name of the wage payroll item you entered previously. See Figure 5-8.

- Be sure you select **Track Expenses By Job** to get workers' comp costs in your job costs reports. Each employee's workers' comp cost will be divided by the jobs the employee worked on during the pay period.

- Click **Next**.

- In the Agency for company-paid liability window, in the first box, enter your workers' comp insurance carrier's name. In our example, it's State Compensation Insurance Fund. See Figure 5-9.

- In the second box, enter the number that identifies you to the agency (e.g. an insurance policy number).

- In the third box, enter your workers' comp liability account. In our example, it's 2240 - Workers' Comp Payable.

- In the fourth box, enter your expense account. In our example, it's 5220 - Workers' Compensation Costs.

- Click **Next**.

- In the Tax tracking type window, since this item is not reported on your tax forms, select **None** from the pull-down list. See Figure 5-10.

- Click **Next**.

90   *Contractor's Guide to QuickBooks Pro*

Figure 5-9
This window shows the details of the payee, the account number, and the correct liability and expense accounts for proper accounting.

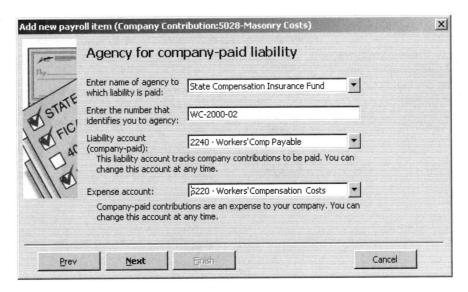

Figure 5-10
Choose the option "None" from the tax classification list for an item you don't report on your tax forms.

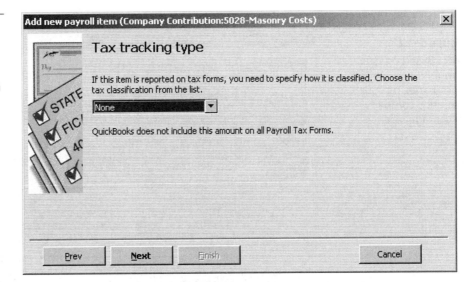

- In the Taxes window, don't select any taxes. Workers' comp company-paid cost isn't subject to any additional federal or state taxes. See Figure 5-11.
- Click **Next**.
- In the Calculate based on quantity window, Select **Calculate this item based on quantity** to multiply each employee's gross wages by the percentage rate your insurance company charges you for the workers' comp classification. See Figure 5-12.

Chapter 5: Payroll Items

Figure 5-11
Leave these items blank because workers' comp company-paid costs aren't taxed.

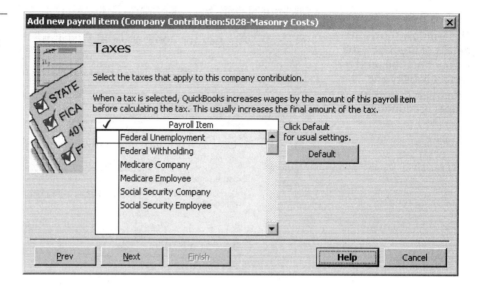

Figure 5-12
Select Based on Quantity to multiply employee gross wages by your insurance company's percentage rate.

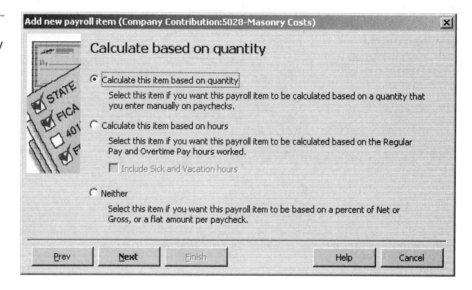

▌ Click **Next**.

▌ In the Default rate and limit window, enter the rate for the workers' comp code in the first box. For example, we entered 9.1%. This is the rate we pay when an employee works on masonry tasks. See Figure 5-13. The rate is given to you by your workers' comp insurance company. The rates for each company contribution will need to be updated when your insurance carrier updates (increases or decreases) your workers' comp rates.

▌ Click **Finish** to complete the new payroll item.

92 *Contractor's Guide to QuickBooks Pro*

Figure 5-13
Enter the correct percentage rate for the workers' comp classification.

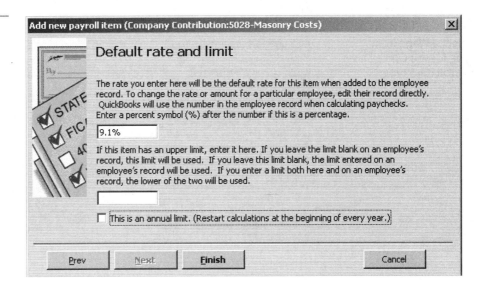

Using Payroll Items If You Don't Track Workers' Comp

If you don't want to track workers' comp, you can add new payroll items or change the existing default payroll items to your own classifications. We suggest you add new payroll items because the default items don't default to the correct account in the Chart of Accounts. We want to make sure field labor is correctly allocated to job costs and office labor is correctly allocated to an overhead expense. We suggest you use these payroll items: Field Labor, Field Labor OT, and Office Labor.

Adding a New Payroll Item

You should enter a new wages payroll item for overhead or office payroll because you'll want to capture the costs associated with your office payroll. To add a new item for Office Labor:

▌ From the **Lists** menu, choose **Payroll Item List**.

▌ In the Payroll Item List window, pull down the **Payroll Item** menu and choose **New**.

▌ Click **Custom Setup** and **Next**.

▌ In the Payroll item type window, select **Wage (Hourly Wages, Annual Salary, Commission, Bonus)**, as shown in Figure 5-14.

▌ Click **Next**.

▌ In the Wages window, select **Hourly Wages**, as shown in Figure 5-15.

▌ Click **Next**.

▌ In the next Wages window, select **Regular Pay**, as shown in Figure 5-16.

Chapter 5: Payroll Items

Figure 5-14
Select a Wage item type for the Office Labor payroll item type.

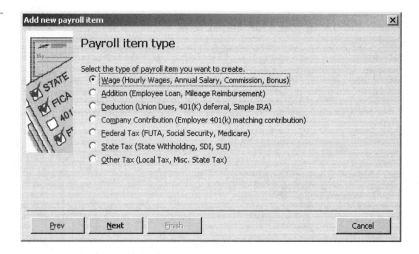

Figure 5-15
Select Hourly Wages for the Office Labor payroll item.

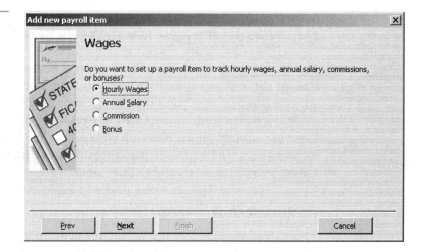

Figure 5-16
Select Regular Pay for the Office Labor payroll item.

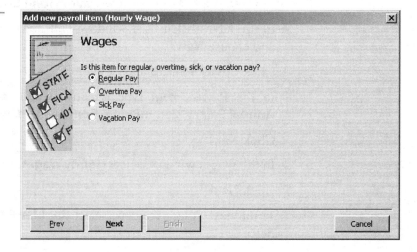

Figure 5-17
Enter the name of the Office Labor payroll item here.

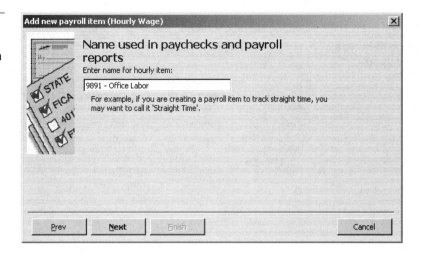

- In the Name used in paychecks and payroll reports window, enter the name of the overhead labor item. In our example shown in Figure 5-17, it's Office Labor.

- Click **Next**. If you type a name that is already a payroll item, a Warning box will appear requesting you use a different name.

- In the Expense account window, enter the account number for tracking this expense. In our example shown in Figure 5-18, it's 6501 - Payroll (office staff). Be sure you select the proper expense account from your Chart of Accounts here. The Expense Account you select here links the payroll item to your Chart of Accounts.

- Click **Finish**.

Figure 5-18
In this example, 6501 - Payroll (office staff) is the correct expense account for the Office Labor payroll item.

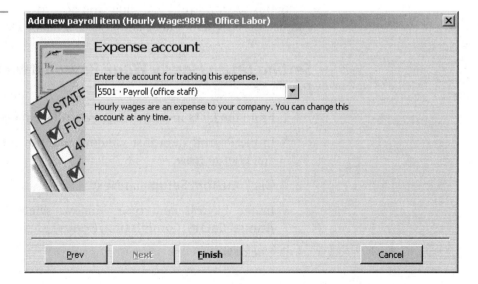

Figure 5-19
Select a Wage item type for the owner/partner payroll item.

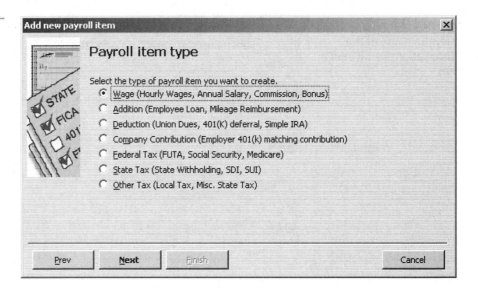

Including Sole Proprietor and Partners' Time Costs in Job Costs

If your company is organized as a sole proprietorship or a partnership, you'll be taking draws instead of issuing payroll checks for yourself or your partner. Because of this, QuickBooks Pro reports won't show any costs for your or your partner's time.

To get these costs into any job cost report, you have to transfer owner or partner time (in dollars) using an hourly wage payroll item and a deduction payroll item. In Chapter 15, Payroll, we'll show you how to actually use the hourly wage item to enter owner or partner time on a timesheet, process payroll, and then deduct the payroll back out of your accounting using the deduction payroll item.

In this section, we'll just show you how to set up the two payroll items you need — Owner's Labor and Owner's Labor Deduction.

Setting Up an Hourly Wage Payroll Item for an Owner or Partner

- From the **Lists** menu, choose **Payroll Item List**.
- In the Payroll Item List window, pull down the **Payroll Item** menu and choose **New**.
- Click **Custom Setup** and **Next**.
- In the Payroll item type window, select **Wage** (**Hourly Wages, Annual Salary, Commission, Bonus**) as shown in Figure 5-19.
- Click **Next**.

96 *Contractor's Guide to QuickBooks Pro*

Figure 5-20
Select Hourly Wages for the owner/partner payroll item.

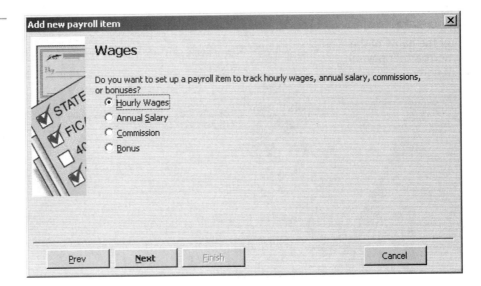

- In the Wages window, select **Hourly Wages** as shown in Figure 5-20.
- Click **Next**.
- In the next Wages window, select **Regular Pay**, as shown in Figure 5-21.
- Click **Next**.
- In the Name used in paychecks and payroll reports window, enter the name of the overhead labor item. In our example shown in Figure 5-22, it's Owner's Labor.
- Click **Next**.

Figure 5-21
Select Regular Pay for the owner/partner payroll item.

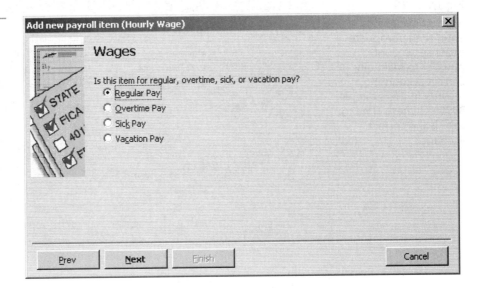

Figure 5-22
Enter the name of the owner/partner payroll item here.

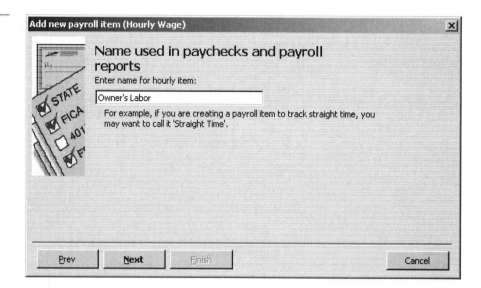

- In the Expense account window, enter the account number for tracking this expense. In our example shown in Figure 5-23, it's 3999 - Owner's Time to Jobs. Be sure you select the proper expense account from your Chart of Accounts here. The Expense Account you select here links the payroll item to your Chart of Accounts.

- Click **Finish**.

Figure 5-23
It's important to enter the appropriate expense account number to maintain correct accounting records. In this example, 3999 - Owner's Time to Jobs is the correct account.

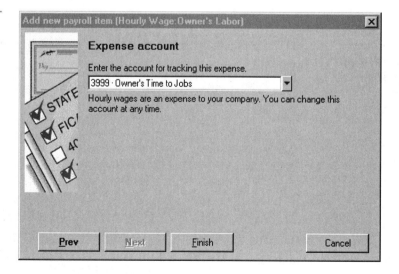

Setting Up a Deduction Payroll Item for an Owner or Partner

When you set up this deduction item, you'll be able to reverse wages and payroll taxes from the owner's check.

- From the **Lists** menu, choose **Payroll Item List**.
- In the Payroll Item List window, pull down the **Payroll Item** menu and choose **New**.
- Click **Custom Setup** and **Next**.
- In the Payroll item type window, select **Deduction**, as shown in Figure 5-24.
- Click **Next**.
- In the Name used in paychecks and payroll reports window, enter the name of the deduction item. In our example shown in Figure 5-25, it's Owner's Labor Deduction.
- Click **Next**.
- In the Agency for employee-paid liability window, leave the first box blank. In the second box, in this case, you wouldn't use a number, just a description of how you're using this payroll item. In our example, we've entered To Deduct Owner's Lab, as shown in Figure 5-26.
- In Liability account, enter 3999 - Owner's Time to Jobs.
- Click **Next**.

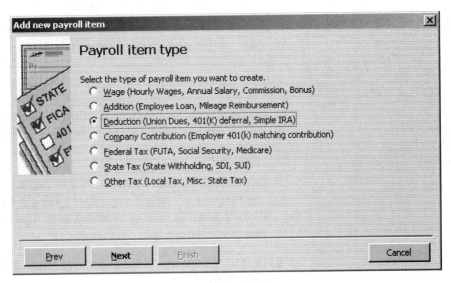

Figure 5-24
You also need to set up a Deduction payroll item type when you set up an owner/partner payroll item.

Chapter 5: Payroll Items

Figure 5-25
Enter the Owner's Labor Deduction in the name field.

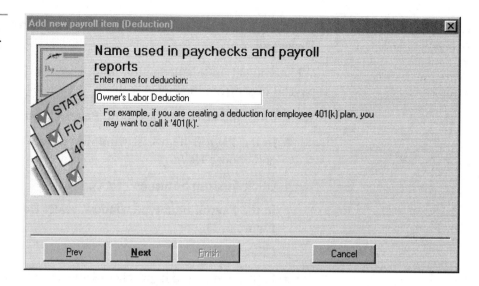

Figure 5-26
Leave the Name field blank, enter a short description in the Number field, and in the Liability Account field enter account number 3999 - Owner's Time to Jobs.

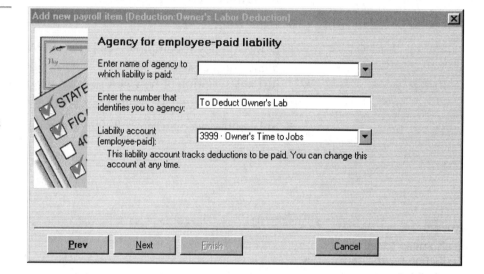

- In the Tax tracking type window, select **Compensation** from the pull-down list, as shown in Figure 5-27. This is important!
- Click **Next**.
- In the Taxes window, because this deduction isn't associated with any taxes, don't select anything. QuickBooks Pro will give you the Payroll Item Taxability warning box. Ignore the warning and click **Yes**. See Figure 5-28.
- In the Calculate based on quantity window, select **Neither**. See Figure 5-29.
- Click **Next**.

100 Contractor's Guide to QuickBooks Pro

Figure 5-27
Select Compensation from the tax classification list for the Owner's Labor deduction payroll item.

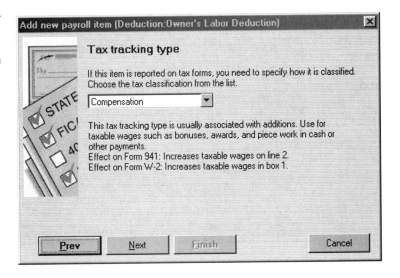

Figure 5-28
Leave the tax boxes blank because there's no tax associated with the Owner's Labor deduction.

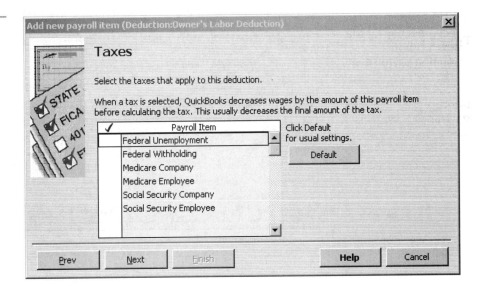

■ In Gross vs. Net, select the gross pay radio button. See Figure 5-30.

■ Click **Next**.

■ In the Default rate and limit window, enter 100% to automatically deduct 100% of the owner's time. See Figure 5-31.

■ Click **Finish**.

Keep in mind that this chapter just showed you how to set up payroll items. Later in the book, in Chapter 15, we'll show you how to use the payroll items you set up here to process a payroll.

After you set up your payroll items, you're ready to move on to setting up classes.

Chapter 5: Payroll Items

Figure 5-29
Select Neither when you set up the deduction payroll item for owner/partner labor.

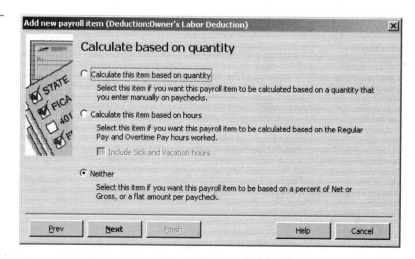

Figure 5-30
Select gross pay for the Owner's Labor Deduction payroll item.

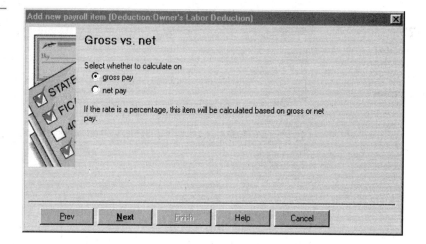

Figure 5-31
Use a default rate of 100% to automatically deduct the owner's time.

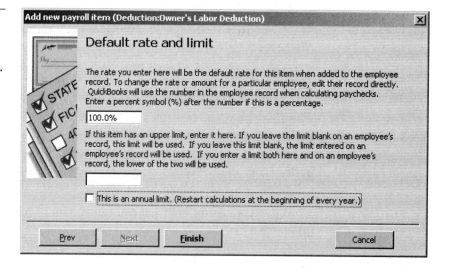

Chapter 6

Classes

Classes are one of the "containers" QuickBooks Pro gives you to earmark transactions and retrieve them later in useful reports. Classes, in fact, are accessible from almost every transaction in QuickBooks Pro. So, when you fill in timecards, you can pick a class to "aim" the time to. Or if you're filling in an invoice, you can pick a class to toss the line item cost into.

Using Classes to Track Cost Categories

We recommend that you use classes to track Labor, Materials, Subcontractors, Equipment Rentals, and Other.

Your tax preparer can also use class reports to get information to prepare your tax return. It doesn't matter whether you're a sole proprietor, partnership, or corporation, your tax return has a section for Cost of Goods Sold. In a construction company, that's a job related cost. The Cost of Goods Sold section on the tax return asks for a summary of your job costs by Labor, Materials, Subcontractors, Equipment Rental, and Other. If you break out your classes into these categories, you can easily produce the reports your tax preparer will need.

We've included a memorized report named "Job Cost Class Report" in the *sample.qbw* file. This report shows total Revenue, Labor, Materials, Subcontractors, and Other subtotaled for the year. Figure 6-1 shows part of a report on the class "Material Costs - Job Related." In Chapter 17 (Reports), we'll show you how to access this report.

Chapter 6: Classes

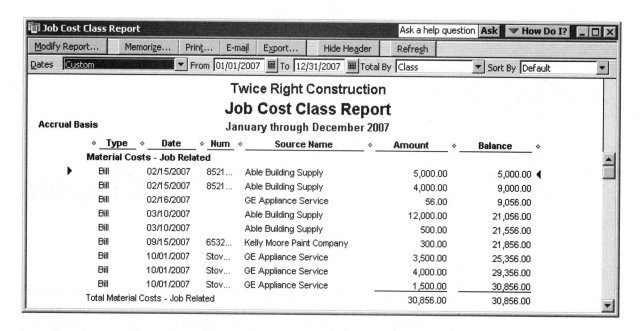

Figure 6-1
Part of a memorized Job Cost Class Report on the class Material Costs - Job Related.

How to Create a Class

Before we tell you how to create a class, you need to make sure the Class Tracking Preference is turned on. To do this:

▌ From the **Edit** menu, choose **Preferences** and click the **Accounting** icon.

▌ On the Company Preferences tab, select **Use class tracking**.

▌ Click **OK**.

Now to create a class:

▌ From the **Lists** menu, choose **Class List**. See Figure 6-2.

▌ In the Class List window, pull down the **Class** menu and choose **New**.

▌ Enter a name for the new class.

If you want more detailed information — a breakdown on labor, perhaps — you can define this new class as a subclass of another class. For instance, you might want to track a specific type of labor, such as trim carpentry, to see if it's coming in as estimated. Just tell your trim carpenters to accurately record the time they spend doing casing and base, and enter the time in a subclass you set up to track that task. Then you can pull up a report

104   Contractor's Guide to QuickBooks Pro

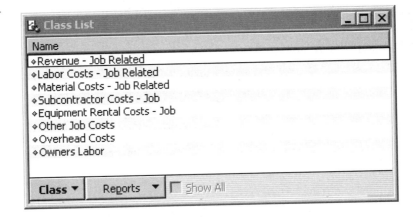

Figure 6-2
You can use the Class List provided in the company.qbw file as a model for your business.

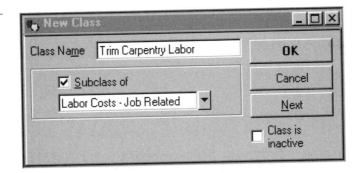

Figure 6-3
You can create subclasses to get a more detailed breakdown of a class.

that calculates the total amount of time/money spent doing casing and base. If you divide the total amount of casing and base installed by the cost of doing it, you'll get a unit cost for placing casing and base. The next time you bid casing and base, you'll have a better unit cost and should be able to make more money on that phase of your business.

To create a subclass as shown in Figure 6-3:

- Click the box in front of **Subclass of** and enter the appropriate class name.

You'll also notice Class is inactive in the New Class window. If you don't want a class to appear in the Class List, you can check this field to keep it from appearing in the list. To make the class active again, click Show All in the Class List window. Inactive classes are marked with a small "x" icon beside the class name. Click this icon to activate the class again.

- Click **OK** to create the new class as a subclass.

After you finish setting up the classes you want to use, move on to the next chapter where we'll show you how to set up customers and jobs.

Chapter 7

Customers and Jobs

QuickBooks Pro tracks jobs by customer — but it's important to understand that a customer can have more than one job. You can set up a main job, then track changes by setting up additional jobs using a change order number. Or you may do more than one job for a customer and want to keep them separate. Then you can assign the jobs different names, such as Rachel Olsen: Bathroom Addition and Rachel Olsen: Guest Cottage. First set up the customer, then add the job or jobs.

How to Set Up a Customer

QuickBooks Pro keeps your customer information in a list called Customer:Job List. Figure 7-1 shows a customer list.

You can use this list to keep track of a customer's billing and shipping address, and business, home, and fax telephone numbers. The billing address might be the existing residence and the shipping address might be the address of the new residence if you're building a new home for the client.

To enter a customer:

- From the **Lists** menu, choose **Customer:Job List**.

- In the Customer:Job List window, pull down the **Customer:Job** menu and choose **New**.

- Enter the customer name. See Figure 7-2. Enter the last name first if you'd like to sort your list alphabetically by last name. Fill in the rest of the fields in the **Address Info** tab.

Chapter 7: Customers and Jobs

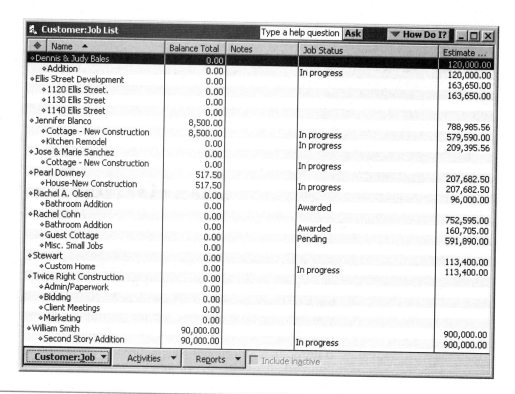

Figure 7-1
The Customer:Job List window displays all your existing customers and their associated jobs.

Figure 7-2
Use the Address Info tab to make additions or changes to a customer's name, contact, phone, or address.

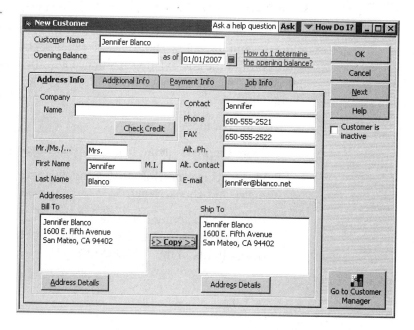

108 Contractor's Guide to QuickBooks Pro

Figure 7-3
The Additional Info tab lets you track more details about a customer. You can even create custom fields specific to your needs here.

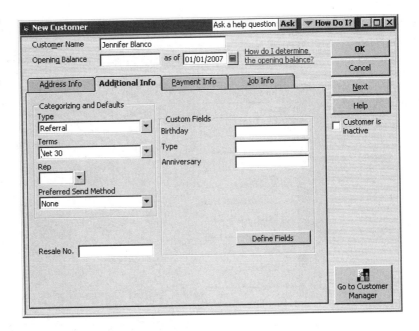

The Additional Info Tab

In the Additional Info tab window, you'll find drop-down lists for Customer Type, Terms, and Rep. You can use Customer Type as a "container" to track customers based on something that's meaningful to you. For example, you could record how the customer got your company's name — newspaper ad, referral, Chamber of Commerce, etc. See Figure 7-3.

- In **Type**, select the method your customer used to get your name (used to track marketing efforts). In our example, we used **Referral**.

- In **Terms**, select the terms of credit you wish to offer the customer and what you want printed on the invoice.

- In **Rep**, you can assign a salesperson to the customer and track sales by representative. If you don't have salespersons, enter the project manager's initials to track who is responsible for the job. The rep's initials can be printed on invoices.

You can also create and use custom fields in this window. You can track information about a customer, such as their birthday or anniversary, so you can do some "target" marketing with a birthday card or gift. You might also use this section to record the type of client. Some contractors track whether a client is "from Hell" or "from Heaven." Naturally, those categorized as "from Hell" are charged more for your efforts.

Figure 7-4
Use the Define Fields window to enter a label for a custom field and specify if it will be used for customers, vendors, or employees.

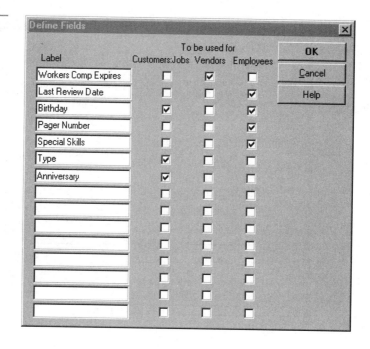

To create a custom field:

- In the Custom Fields area, click **Define Fields**. See Figure 7-4.

- In the Define Fields window, under **Label**, enter the field names you want to create.

- In the columns to the right, check the boxes to select where you'll be using this field (Customers:Jobs, Vendors, Employees).

- Click **OK** to create the new fields.

The Payment Info Tab

You can enter customer payment information in the Payment Info tab window. See Figure 7-5. If you require your customers to pay by credit card (very rare in construction), you can enter the customer's credit card information on this screen. If you enter your customer's credit card number and expiration date in QuickBooks Pro, you should immediately set up password protection and make sure that only a trusted few have access to viewing customer payment information.

- Click the **Payment Info** tab of the New Customer window to fill in more information about a customer.

- In **Account**, enter the account number or the PO number your customer requested you put on each invoice you send to them.

- We suggest that you skip the Job Info tab when you're entering information for a new customer. We'll cover it in the next section.

- Click **OK** to enter the new customer.

Figure 7-5
Enter customer payment information in the Payment Info tab window.

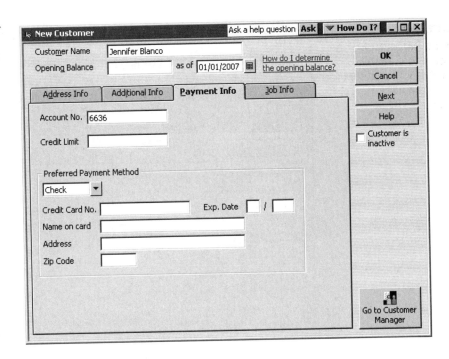

Adding a Job for a Customer

Whether you do one or more jobs for a customer, it's a good idea to set up each job separately. For example, if Twice Right adds a bathroom for Rachel Olsen, and then later builds her a guest cottage, both the bathroom addition and the guest cottage are separate jobs, but they should be attached to the same customer — Rachel Olsen. In our example, we've entered a Guest Cottage job for our customer Rachel Olsen. To add a job for a customer:

- From the **Lists** menu, choose **Customer:Job List**.
- In the Customer:Job List window, click the customer you want to add a job for.
- From the **Customer: Job** pull-down menu, choose **Add Job**.
- Enter a job name in the box next to **Job Name** on the New Job window. In our example, shown in Figure 7-6, we entered Guest Cottage.
- Change information on the **Address Info** and **Additional Info** tabs if you need to.
- Click the **Job Info** tab.

Figure 7-6
Use the Job Info tab to record all the pertinent information for a particular job.

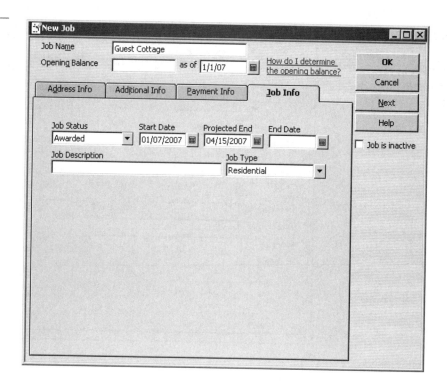

The Job Info Tab

You can enter Job Status, Start Date, Projected End, End Date, Job Description, and Job Type on the Job Info tab.

The Job Status box lets you track the status of your jobs. Pending, Awarded, In progress, Closed, Not awarded, and None are your choices. You can do a report on jobs Pending, or jobs Not awarded. A report on Not awarded vs. all your jobs will tell you the ratio of jobs you're getting to the jobs you're bidding. Use this information when you create your marketing budget. Figure 7-6 shows a Job Info tab window.

- From the pull-down Job Status list, choose the status of the job.
- In **Start Date**, enter the start date for the job.
- In **Projected End**, enter the projected end date for the job.
- In **End Date**, enter the actual date the job ended. (You'll need to come back later and fill this in.)
- In **Job Description**, enter the description for the job. You would fill this in if you use job numbers instead of names in the Job Name field.

The Job Type box in the Job Info tab window has a drop-down list from which you can choose a job type. For example, you could set up a job type such as Residential, Commercial, or Remodel, with subtypes for Contract or Speculation. Then, whenever you get a new job, you can choose which

Figure 7-7
You specify a name for a new job type in the New Job Type window.

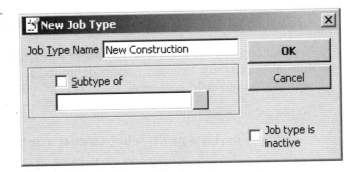

Figure 7-8
Use subtypes to get a detailed breakdown of a job type.

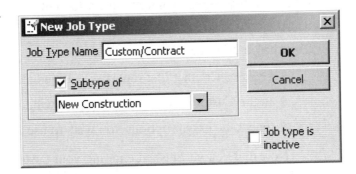

job type the job falls under. Later (at year's end, for example) you can sort jobs by type to make a report that evaluates profitability by job type. Information like this can help you decide whether or not to continue to do one type of work. If your commercial jobs aren't as profitable as your residential work, you might decide to stop bidding commercial jobs.

Setting Up a New Job Type

To set up a new job type:

- From the drop down **Job Type** list in the Job Info tab window, choose **Add New**.

- In the New Job Type window, in **Job Type Name**, enter the new job type name. See Figure 7-7.

- If you want the new job type to be a subtype of another job type, click **Subtype of** and enter the appropriate job type. See Figure 7-8.

- Click **OK** to enter the new job type and return to the New Job window.

- Click **OK** in the New Job window to add the job for the customer.

Exporting or Printing Your Customer List

You can export your Customer:Job list to a program such as Excel, ACT, or any other program that can import delimited text. You can also print out reports for your field personnel.

To run a customer report:

- From the **Lists** menu, select **Customer:Job List**.
- Click on the **Reports** pull down menu and select **Contact List**. In the Modify Report window, select the columns you want to display on your report. Click **OK**.

To export a customer list:

- Create a Contact List report as outlined above.
- In the Customer Contact List report window, click on **Export**.
- Select the export options you wish to use and click on **Export**.

Now that you've finished entering your customers and jobs, let's move on to Chapter 8 to set up your vendors and subcontractors.

Chapter 8

Vendors and Subcontractors

In QuickBooks Pro, there's no difference between a subcontractor and a vendor. Whether they supply you with materials, services, or both, you set up vendors and subs the same way. But how you set them up will determine how you can create your reports on them. It's important to closely follow our instructions for setting up vendors, especially when you set up 1099 vendors.

Setting Up 1099 Vendors

You're required to report on IRS Form 1099 the yearly total your company paid each vendor that supplied over $600 of labor. If you send 1099-MISC forms to any vendors or subcontractors, you can set up QuickBooks Pro to track all 1099-related payments to each vendor. Then at the end of the year, QuickBooks Pro can print your 1099-MISC forms.

To set up QuickBooks Pro to track 1099s, you need to set the Tax:1099 preference.

- From the **Edit** menu, choose **Preferences**.
- Click the **Tax:1099** icon.
- In the Company Preferences window, click **Yes** to answer **Do you file 1099-MISC forms?**

Next you need to associate the appropriate accounts with each 1099 category you use to report to the IRS. Most construction businesses report amounts only for Box 7: Nonemployee Compensation.

- Click in the Account column next to **Box 7: Nonemployee Compensation**.

Chapter 8: Vendors and Subcontractors

Figure 8-1
The Vendor List window displays each vendor you have entered into QuickBooks Pro and the amount you owe each vendor.

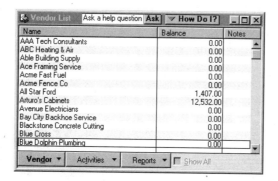

▌ In the drop-down Account list, select the account you've been using to track 1099 payments to your vendors. If you've used more than one account, click **Selected accounts** at the top of the list, and then select each individual account.

Remember that each 1099 category has to have its own unique account. If you report Nonemployee Compensation (Box 7), you could use your Cost of Goods Sold account named Job Related Costs to track Nonemployee Compensation. But once you select Job Related Costs for Box 7, you can't use it to track any of the other 1099 categories.

▌ Click **OK** to save your preference changes.

Now you're ready to create a new 1099 vendor.

▌ From the **Lists** menu, choose **Vendor List**. See Figure 8-1.

▌ In the Vendor List window, pull down the **Vendor** menu and choose **New** to enter a new vendor. Figures 8-2 and 8-3 show New Vendor windows for Eric Savage.

Figure 8-2
Use the Address Info tab to add or change a vendor's name, contact, phone, or address.

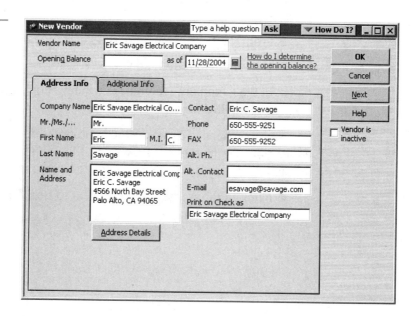

116   *Contractor's Guide to QuickBooks Pro*

Figure 8-3
Use the Additional Info tab to track more details about a vendor. You can also create custom fields specific to your needs here.

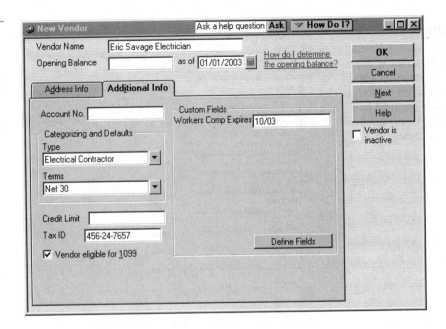

- Enter the vendor's name and address.

- In **Company Name**, enter the name as you want it on your QuickBooks Pro Vendor list. This name won't appear on the 1099 form if you also fill in **First Name**, **M.I.**, and **Last Name**.

If the vendor is a sole proprietor, you have to print the person's name, not the company name, on the 1099. Because of this, you'll need to enter the person's legal name in the First Name, M.I., and Last Name fields. If the vendor's ownership is classified as a partnership, don't fill in the First Name, M.I., or Last Name fields (you can enter a contact name in the contact field). Enter the legal partnership name in the Company Name field. If the vendor is incorporated you don't have to send them a 1099.

If part of the address in the Address field includes a name on line 2 or below, be sure the name is identical to the Company Name field or the First Name, M.I., and Last Name fields. Otherwise, it will print as part of the address on the 1099 form.

- Be sure you enter the company's two-letter state abbreviation and zip code. Click the **Address Details** button to enter this information.

- Click the **Additional Info** tab. See Figure 8-3.

- Fill in **Tax ID**. If the vendor is a sole proprietor, enter the vendor's Social Security Number or EIN (employer identification number).

Figure 8-4
Use the Define Fields window to enter a label for a custom field and specify if the custom field will be used for customers, vendors, or employees.

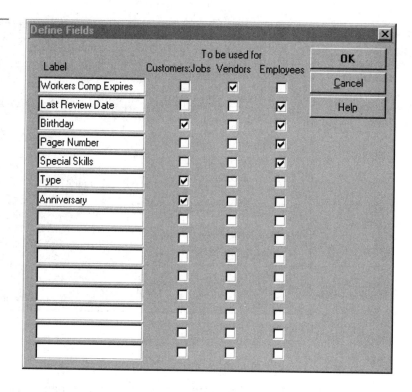

- Select **Vendor eligible for 1099**.

- Fill in **Account No.** if you wish. This information will be printed in the Memo field of any checks to the vendor. If you have an account number with a vendor, you can enter it here and it'll automatically be printed on each check to the vendor.

You can set up custom fields in the Additional Info window. One field you might want to add is a custom field called Workers Comp Expires. You can use this field to track the expiration date of the workers' comp insurance for each subcontractor.

To create a custom field:

- In the Custom Fields area of the Additional Info tab window, click **Define Fields**. See Figure 8-3.

- In the Define Fields window, under **Label**, enter the field names you'd like to create. See Figure 8-4.

- In the columns to the right, check the boxes to select whether the label is used for customers, vendors, or employees.
- Click **OK** to create the new labels.

Now you're ready to save all information about the vendor:

- Click **OK**.

Setting Up a Non-1099 Vendor

You set up a non-1099 vendor the same as a 1099 vendor, with these two exceptions: First, you don't have to enter a tax ID number. Second, you don't select Vendor eligible for 1099 in the New Vendor Additional Info tab window. For Vendor Type, enter the appropriate type such as HVAC contractor, electrical subcontractor, and so on.

Now that you've finished setting up your vendors, let's set up your employees. If you don't have employees, skip Chapter 9 and move on to Chapter 10 where we will cover Opening Balances.

Chapter 9

Employees

There are two cases when you'll want to set up your employees in QuickBooks Pro using the Employee List. First, use it if your company has employees and you process your own payroll. Second, use it if you want to track employee time by job even though your payroll is processed by a service.

Setting Up Your Employee List

The Employee List is a listing of every employee record you enter into QuickBooks Pro. Each employee's record contains all the information you need to calculate payroll, whether the employee is on monthly salary or getting a weekly paycheck based on hours worked.

The Payroll module in QuickBooks Pro is very flexible. You can use it to figure out all common payroll items such as federal taxes, FUTA, Social Security, Medicare, state unemployment insurance, and state disability. Naturally, you can track, calculate, and print W2s, 1099s, and Forms 940 and 941. You can also track sick time and vacation pay and create your own custom additions and deductions.

It's easy to add, edit, or delete employees from the Employee List. To add employees, use the employee template you set up in the Payroll and Employees preference in the Employee Defaults window. The template contains the information common to all employees, so you only have to enter individual employee information to set up a new employee. If an employee goes out on disability or maternity leave, you can assign him/her as "Inactive" in the Employee List.

To set up an employee:

▌ From the **Lists** menu, choose **Employee List**. See Figure 9-1.

Chapter 9: Employees

Figure 9-1
The Employee List window displays the Social Security number and the name of every employee you have entered into QuickBooks Pro.

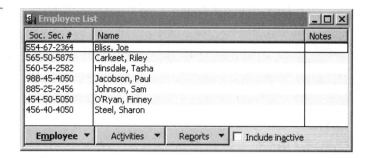

Figure 9-2
Use the Personal tab to add or change an employee's name, Social Security number, etc.

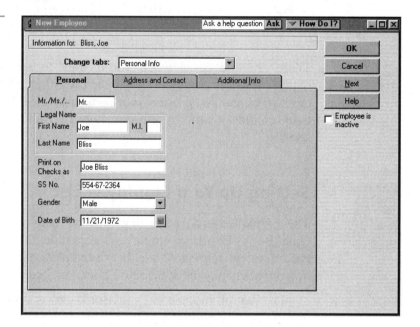

■ In the Employee List window, pull down the **Employee** drop-down menu and choose **New** to enter a new employee. The figures in this chapter show how we entered new employee Joe Bliss. See Figure 9-2. You can use them as an example for entering your employees.

Using the Personal Tab

■ Fill in **Mr./Ms.**, **First Name**, **M.I.**, and **Last Name**.

■ In **Print on Checks as**, enter the name as you want it to appear on the employee's paycheck and W2.

■ In **SS No.**, be sure you enter the employee's Social Security number correctly. It's used on quarterly reports and W2s.

■ The **Gender** and **Date of Birth** fields are for reference only. You use them to store employee information.

Using the Address and Contact Tab

- Be sure you enter the employee's address, city, two-letter state abbreviation and zip code.

- **Phone**, **Cellular**, **Alt. Phone**, **Fax**, **E-mail** and **Pager** fields are also for reference only. Use them to store employee information.

Using the Additional Info Tab

Use **Custom Fields** to create new fields to collect additional data for all your employee records. QuickBooks Pro treats the information you enter in a custom field the same way it treats information entered into any other field. If you export a list that contains data in custom fields, QuickBooks Pro exports that data along with the other data from the list. Then you can sort the exported field and use it in reports that QuickBooks Pro doesn't provide.

Here are examples of some employee custom fields you can use:

- last review date
- birthday
- pager number
- productivity numbers
- goals
- special skills

To create a custom field:

- In the New Employee window, click the **Additional Info** tab. See Figure 9-3.
- In the Custom Fields area, click **Define Fields**. See Figure 9-3.
- In the Define Fields window, under **Label**, enter the field names you want to create. See Figure 9-4.
- In the columns to the right, check the box for **Employees**.
- Click **OK** to create the new fields.

Using the Payroll Info Tab

Using the Payroll Info tab window, you can set up information for each employee on:

- earnings
- additions, deductions, and company contributions
- pay period

Chapter 9: Employees

Figure 9-3
Use the Additional Info tab to track more details about an employee. Create custom fields you need here.

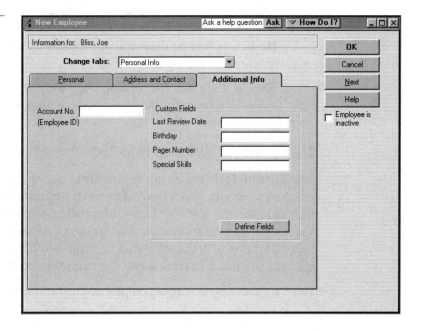

Figure 9-4
Use the Define Fields window to enter a label for a custom field and specify if it will be used for customers, vendors, or employees.

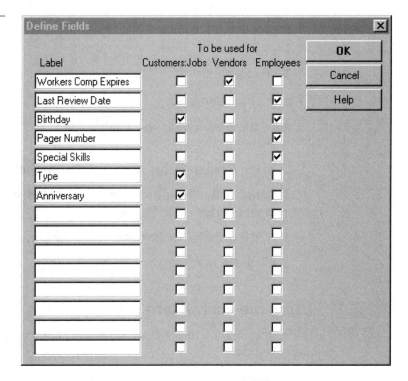

124   *Contractor's Guide to QuickBooks Pro*

- class
- taxes
- sick/vacation
- direct deposit

To get to the Payroll Info tab window:

- In the New Employee window, from the **Change tabs** drop-down list, select Payroll and Compensation Info. See Figure 9-5.

- Under Earnings, select all payroll items from the drop-down list that apply to the employee. (See Chapter 5 if you haven't set up payroll items.) For example, the employee in our example works on concrete and/or masonry and painting so his payroll information includes the concrete, masonry and painting payroll items. The payroll item determines what account the payroll expense is posted to, such as to 5110 Job Related Costs.

- Select an **Hourly/Annual Rate** for the employee. Realize that the payroll item determines where the payroll expense is posted, and the hourly/annual rates are the actual base wage being used in calculating the payroll as well as the job costing. For Hourly, enter in the hourly rate. For Salary, enter annual salary. You're only allowed eight payroll items per employee.

- Select the **Use time data to create paychecks** box if you plan on using the QuickBooks Pro timecard to enter the employee's time. If you select this button, the timecard information entered will flow into the paycheck for this employee.

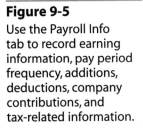

Figure 9-5
Use the Payroll Info tab to record earning information, pay period frequency, additions, deductions, company contributions, and tax-related information.

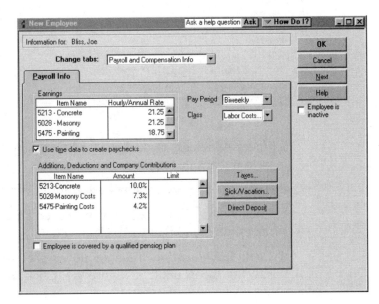

Figure 9-6
Use the Federal, State and Other tabs to specify tax information for each employee.

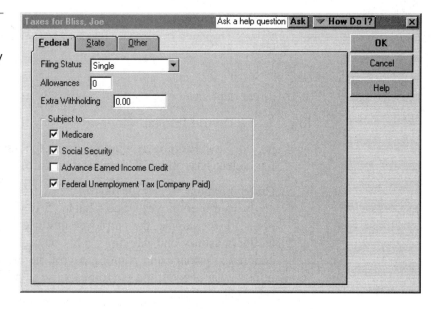

Figure 9-7
For eligible employees, use the Sick & Vacation window to track available hours and accrual information.

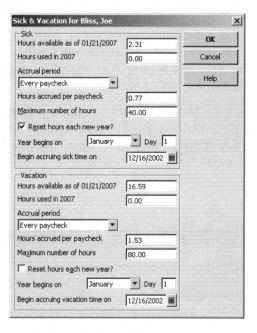

- Under Additions, Deductions, and Company Contributions, select all payroll items that apply to the employee.
- Select the appropriate **Pay Period** for the employee.
- Click **Taxes** and make sure you select all the appropriate boxes. See the example in Figure 9-6. Click **OK**.
- Click **Sick/Vacation** if the employee is eligible for sick or vacation pay. See the example in Figure 9-7. Click **OK**.

▮ If you've subscribed to QuickBooks Pro's payroll service, click **Direct Deposit** to set up direct deposit of the employee's pay if appropriate. Click **OK**.

Now you're ready to save all information about the employee:

▮ In the New Employee window, click **OK**.

Setting up a Sole Proprietor/Partner (Employee) Card

In Chapter 5 we discussed how to include a sole proprietor or partner's time in job cost reports. In this section, we'll walk you through setting up the employee card for the sole proprietor or partner. Keep in mind that a sole proprietor or partner (non-incorporated business) does not get paid through payroll, instead they take money out of the business in the form of a withdrawal. You will set up an employee card if the owner works on jobs and wants to track the costs. In Chapter 15 we'll walk you through processing a zero payroll check to the owner (in order to job cost the owner's time to a job).

To set up a sole proprietor/partner employee card:

▮ From the **Lists** menu, choose **Employee List**. See Figure 9-1.

▮ From the **Employee** drop-down list, select **New**.

▮ Enter the Employee information on the Personal, Address and Contact, and Additional Info tabs. See Figures 9-2 and 9-3.

▮ In **Change Tabs**, select **Payroll and Compensation Info**. See Figure 9-8.

▮ In the **Earnings** box, select Owner's Labor (if you don't see it in the list refer back to Chapter 5, Payroll Items).

▮ In **Hourly/Annual** rate, enter the amount you would pay someone else with the same qualifications to do the work the owner performs on the job. The rate should be per hour including labor burden. Labor burden is the cost of payroll taxes, workers' comp and employee benefits. Do not enter in the amount the owner withdraws; instead you should enter the amount you would have to pay someone else to do that job. Later, when you are analyzing your job costs reports, you'll want to know which jobs are profitable, including all the time the owner put into the job.

▮ In **Class**, select Owner's Labor. If it doesn't appear in your list, you should add it now. See Chapter 6 for more information about classes.

▮ Click the button next to **Use Time Data to Create Paychecks**.

▮ Under Additions, Deductions and Company Contributions, enter Owner's Labor Deduction (if it doesn't appear in your list, see Chapter 5 to learn how to enter it). In amount, enter 100%. It will automatically appear as negative 100% (or –100%). You'll learn more about why this is important in Chapter 15.

▌ Click on **OK** to save the employee information.

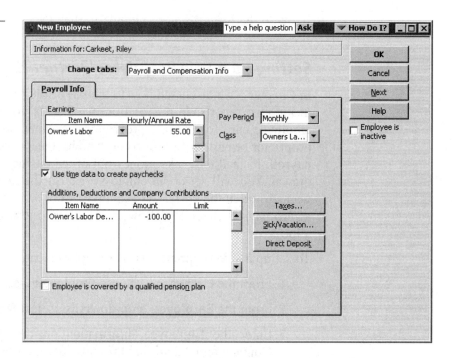

Figure 9-8
Set up payroll information to record job costs for a sole proprietor or partner.

Now that you've finished setting up customers, jobs, vendors, and employees, it's time to tell QuickBooks Pro what your opening balance is for each account you have. Chapter 10, Opening Balances will show you how to enter those balances.

Chapter 10

Opening Balances

Getting your computerized accounting system into balance can be a trying experience. If you're already using a computerized accounting system, you should make sure your balances are accurate before transferring them over to QuickBooks Pro.

Keep in mind when you begin using QuickBooks Pro that you don't have to start with opening balances and can always go back and change your beginning balances.

Entering Opening Balances in QuickBooks Pro

A journal entry is used to get your beginning balances into QuickBooks. If you don't know what your beginning balances are, your accountant or tax preparer can help you arrive at those numbers.

- From the **Company** menu, choose **Make General Journal Entries** to get the General Journal entry window.

- Enter **Asset**, **Liability**, and **Capital** account balances. Keep in mind that Assets usually have debit balances (except accumulated depreciation), while Liabilities and Capital have credit balances. See Figure 10-1.

> ***Note:*** *If you're starting to use QuickBooks Pro at the beginning of your fiscal year, you enter only Asset, Liability, and Capital account balances. Don't enter beginning income, cost of goods sold, or expense account balances.*

When you finish, print a Balance Sheet to make sure everything balances. To print a Balance Sheet:

- From the **Lists** menu, choose **Chart of Accounts**.

- In the Chart of Accounts window, pull down the **Reports** menu, choose **Report on All Accounts**, then **Balance Sheet**, and then **Standard**. The date of the report should be the date you started using QuickBooks Pro.

> *Note:* Don't enter Accounts Receivable or Accounts Payable balances. They'll come in automatically when you enter the outstanding invoices (receivables) and outstanding bills (payables).

Entering Invoices for Accounts Receivable

Accounts Receivable is money your customers owe you as of the date you start using QuickBooks Pro to enter transactions. If you've been using another accounting system, you should be able to print out a report with each customer's name, date of each outstanding invoice, invoice number, and the amount due and transfer the information to QuickBooks Pro manually.

Figure 10-1
This is an example journal entry to get beginning balances into QuickBooks.

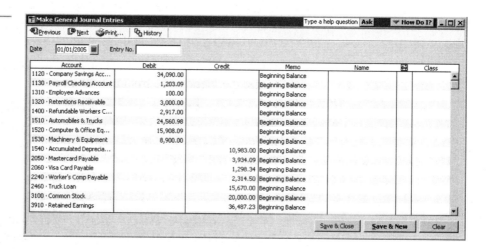

Enter each outstanding Accounts Receivable invoice as an Intuit Service Invoice with the date of the original invoice, *not* the date you enter the invoice. You won't be sending out these invoices — they're simply a way to enter your Accounts Receivable balances. In Figure 10-2, we've entered a beginning balance for Jennifer Blanco.

To enter an outstanding customer invoice:

- From the **Customers** menu, choose **Create Invoices**.
- In **Template**, select **Intuit Service Invoice**.
- In **Customer:Job**, select the customer's name and job. If you entered an estimate for this job, you'll see the Available Estimates dialog box. Click **Cancel** to close the dialog box.
- In **Date**, enter the original date of the invoice, not the current date.
- In **Invoice #**, enter the original invoice number.

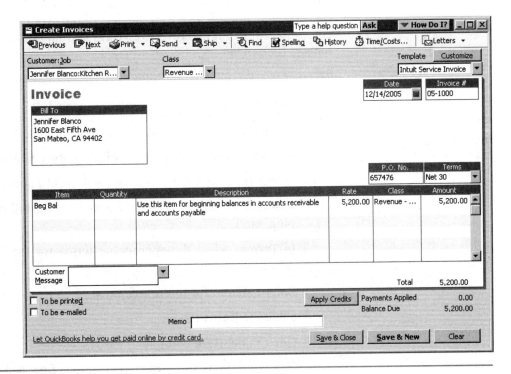

Figure 10-2
Use the Create Invoices window to enter each outstanding invoice to make sure your Accounts Receivable balance is correct.

- In **P.O. No.**, enter the purchase order number for the invoice or job if the customer gave you one.
- In **Terms**, select the appropriate terms for the invoice from the drop-down list.
- Click in the **Item** column.
- From the pull-down Item list, select **Beg Bal**.
- In **Amount**, enter the amount of the invoice.
- Click **Save & Close**.

Entering Bills for Accounts Payable

Accounts Payable is money you owe to vendors (and subcontractors) as of the date you start entering transactions into QuickBooks Pro. If you've been using another accounting system, you should be able to print out a report with the vendor's name, date of each outstanding bill, bill number, and the amount due and transfer the information to QuickBooks Pro. Enter each outstanding Accounts Payable bill with the date of the original bill, *not* the date you enter the bill. In Figure 10-3, we've entered an outstanding bill from Eric Savage Electrician.

To enter vendor outstanding bills:

- From the **Vendors** menu, choose **Enter Bills**.
- In **Vendor**, from the pull-down list, select the vendor's name.
- In **Date**, enter the date of the bill, *not* the current date.
- In **Ref. No.**, enter the vendor's invoice/bill number.
- In **Amount Due**, enter the full amount of the bill (including tax, shipping, etc.).
- In **Terms**, select the appropriate terms for the bill from the drop-down list.
- Enter a memo for future reference.
- Click the **Items** tab.
- Click in the **Item** column.
- From the Item pull-down list, select **Beg Bal**. Enter **Qty**, **Cost**, **Amount**, **Customer:Job**, and **Class**.
- Click **Save & Close**.

Figure 10-3
Use the Enter Bills window to enter each outstanding bill to make sure your Accounts Payable balance is correct.

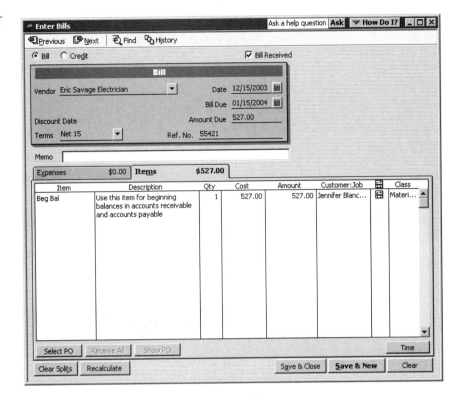

Now you've set up opening balances in QuickBooks Pro and you should be ready to enter new transactions in your company's data file. Remember though, you can always go back and make changes to the file if you find that something is out of balance.

Chapter 11

Organizing Work Flow

Even with the latest computer hardware and all the power of QuickBooks Pro, you still may not have all the data you need. Why? Because to get the most from your system, you need to have an organized flow of information from the field, through the office, to your computer system. For QuickBooks Pro to do its job you'll need to study — and possibly change — how work flows at your business.

For example, QuickBooks Pro has a flexible and powerful payroll capability. It's especially useful for allocating the cost of labor to multiple jobs. But unless you get that payroll information into your computer system, you won't be any better off than before you started using QuickBooks Pro.

Whether you work alone or have office help, office and paperwork flow should follow a strict process. If you don't have one consistent written method for organizing paperwork, you may find yourself wasting many valuable hours shuffling papers rather than generating income for your business.

The best way to get a new routine organized is to get everyone involved. For example, before you start using a new payroll timecard or a new process for turning in paperwork, plan a lunch meeting with all your field workers. During lunch give them something new that will help them stay organized, such as a binder with their new timecards in it, or a clipboard to keep receipts together until they get back to the office. Treating them to lunch and providing free organizing materials may help motivate them to follow your new procedures.

In our opinion, the more informed you keep the people actually doing the work, the more effective they can be. Usually job cost information needs to make a complete circle so the people creating it get feedback on how they're doing. We recommend that you show your project managers a job cost report once a week. If they understand how you're using the data they provide, they'll be more likely to make sure you get it.

The Job Estimates vs. Actuals Detail is a good report to give your project manager on a weekly basis. It shows a job, broken out by phase, with its estimate, actual, and variance figures. These numbers will help the project manager see where his job is running over, or under, the estimates — while he can still do something about it. Another useful report for project managers is the Vendor Contact List Report. This report includes company name, contact name, phone number, and fax number.

Setting Up Your Office Files

One of the hardest jobs in a construction office is keeping the files current. You need an effective filing system to keep things from piling up. Here's the system we recommend.

For your financial files, let QuickBooks Pro be your guide. Organize your file drawers the same way QuickBooks Pro organizes its lists — by vendors, employees, and so on. It makes good sense to take advantage of the correlation between the QuickBooks Pro lists and your paper files. Intuit spent many hours and dollars figuring out the most efficient system for filing with a computerized system. Put their expertise to work for you, and use this same system for your paper files.

Customers and Jobs Files

You'll need a file drawer, or a section of a file drawer, for customers, with separate folders for each job that you do for that customer. For example, Rachel Olsen originally contacted us about remodeling her kitchen, and at Twice Right we have one folder labeled "Olsen, Rachel — Kitchen Remodel." Then when she decided to build a guest cottage, we started another file labeled "Olsen, Rachel — Guest Cottage." We filed all correspondence to and from Rachel in one of those folders.

Your customer folder should contain:

- any correspondence to or from the customer about the job.
- the original project contract.
- copies of the billings. If the billings are bulky, use a separate customer job file for billing information.
- any change orders you issue.

Your customer folder should *not* contain:

- vendor invoices for the job.
- any documents that don't relate to the job.

Figures 11-1 and 11-2 are flow charts showing recommended job and customer transaction procedures.

136   *Contractor's Guide to QuickBooks Pro*

Chapter 11: Organizing Work Flow

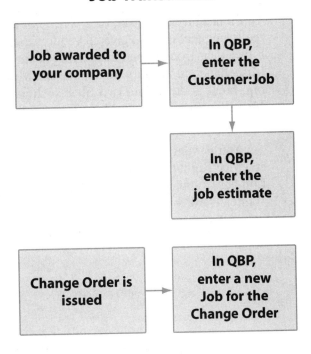

Figure 11-1
This flow chart shows how a job is processed. To see your jobs in QuickBooks Pro, pull down the Lists menu and choose Customer:Job.

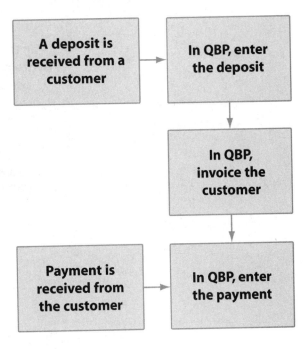

Figure 11-2
This flow chart shows how things you enter in QuickBooks Pro affect a customer.

Vendor Files

You'll need a file drawer, or a section of a file drawer, for vendors. But it's really not necessary to give all vendors their own files. If you only get a few bills a year from a vendor, you may want to file that vendor's invoices in a Miscellaneous Vendors file. When the Miscellaneous Vendors file gets too big, break it down into two files, such as Miscellaneous Vendors A-L and Miscellaneous Vendors M-Z. File all bills in the vendor files whether they're job related or not.

You'll want to set up an individual file for some vendors. For example, at Twice Right Construction we have a charge account with Truitt & White Lumber Company. Because we get several invoices a month from Truitt & White, Truitt & White has its own file.

The flow chart in Figure 11-3 shows how to organize vendor transactions.

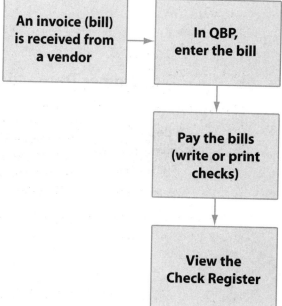

Figure 11-3
This flow chart follows the path of a vendor bill from the time it's received until the bill is paid.

Employee Files

You'll also want to keep a file drawer, or a section of a file drawer, for employees. Since this information is confidential, be sure to keep these files in a locked file cabinet.

Your employee folder should contain:

- a completed employment application.
- a completed I-9.
- the employee's resume, if you have one.
- employee review documents.
- benefits documents.

Organizing Your Payroll

One way to make sure you report labor accurately, on a per job basis, is to use items that describe the phases of labor. See Chapter 4 if you need more information on these items. You can print these items on the back of timecards so employees can use them when they fill out the timecards. Show your employees how using the items helps the business determine true job costs and helps raise profits (and wages). They'll be more likely to fill them out carefully if they understand their importance.

Once the timecards are filled out, it's a simple matter of entering the items into a timesheet in QuickBooks Pro. If you make the items on the back of the timecard match the names of the items in QuickBooks Pro, then entering the timesheet in QuickBooks Pro will simply be a matter of selecting the matching numbers. For example, suppose Joe Bliss has completed 40 hours of carpentry work in a week. His paper timecard would look something like Figure 11-4. In QuickBooks Pro, the timecard would look like Figure 11-5. The "13" item in the Task # column of the timecard matches the "13 Windows" item in the Service Item column on the Timesheet.

If you use computers at your job sites, you can fill in electronic timecards or summary sheets and e-mail them as attachments to the home office.

Keeping Office Paperwork Current

The most important thing is getting the field paperwork into the office. You'll have to start by working with your field people on how, and when, you want them to turn paperwork into the office. Don't get discouraged. When you first start out, you'll probably get all the packing slips, invoices,

Chapter 11: Organizing Work Flow

Date:	1/5/2007								
Employee:	Joe Bliss								
Pay period ending:	1/7/2007								
Customer/Job	Task #	Description	Mon	Tue	Wed	Thur	Fri	Sat	Sun	Total
Blanco/Kitchen	13	Windows	8	8	8					24
Blanco/Kitchen	21	Cabinets				8	8			16
Daily totals:			8	8	8	8	8			40

Figure 11-4
A typical paper timecard computed for a week.

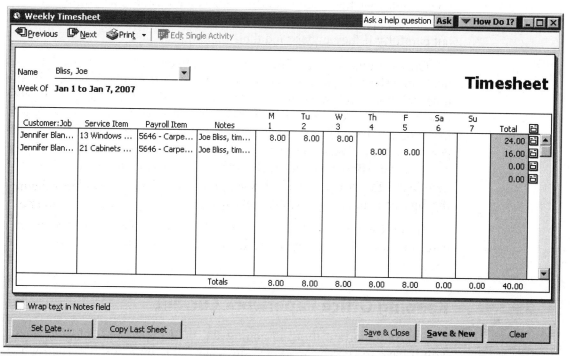

Figure 11-5
A QuickBooks Pro timesheet looks similar to a paper timecard. Notice how the job and item entered match the job and item on the paper timecard.

140   Contractor's Guide to QuickBooks Pro

and receipts turned in torn and crumpled, or not turned in at all, with excuses for why they were lost. But with encouragement, your field people will get the hang of it and start turning in their paperwork on a regular basis.

You'll need a place in the office, such as a basket or box, to use as an inbox for all field and office paperwork that comes into the office. Make sure the field staff can see and reach this inbox easily when they come in to drop off their paperwork.

Sort the paperwork that finds its way to the office. Put payroll timecards in a "To be approved" file. Put vendor invoices in a file for coding and approval.

Enter approved timecards and vendor invoices in QuickBooks Pro. Then move them to the "To be paid 1 - 15th" file or the "To be paid 16th - EOM" file. After you pay an invoice or timecard, file it in the vendor file or employee file.

Another handy job-costing tip is to ask your materials suppliers and subcontractors to bill you by job. Most suppliers won't have any trouble issuing a separate invoice (or at least a separate line item) with a specific job name. Usually you can specify that no materials can be purchased without a job name being assigned to the invoice. As a check, you can also ask your field personnel to save shipping slips and bills of lading and to write the job name on them.

Change Orders

Whether you issue change orders in the office or the field, you need to set up a process to handle them consistently and smoothly. It's important that you inform the office as soon as possible when there's a change order. Change orders can have a major impact on the profitability of a job. The sooner you get them into your system and begin job costing, the sooner you'll know the impact of the changes.

In the first eleven chapters of this book, we concentrated on showing you how to set up QuickBooks Pro for a construction business. In the next chapter, we'll show you how to enter an estimate in QuickBooks Pro.

Chapter 12

Estimating

In this chapter we'll show you how to bring outside summary estimate totals into QuickBooks Pro. We realize that estimating is an individual thing. If your style of estimating is working for you, keep using it. Or if you already use, and are happy with, a sophisticated estimating program that returns useful information that QuickBooks Pro doesn't track, stick with it. You can use a spreadsheet program, or Craftsman's National Estimator, or any of the other estimating programs on the market. As long as your estimating program can create a summary by phase report, you can use it to generate an estimate to hand-enter into QuickBooks Pro.

You can enter the summary information into QuickBooks Pro and track it against your actual costs. Then you can generate a report called Job Estimates vs. Actuals Detail. This report is the most useful report you can get for running your jobs on a day-to-day basis.

QuickBooks Pro has no labor and material cost database. If you use 100 or fewer different items, you may be perfectly happy to enter those 100 items into the QuickBooks Pro Item list. When your costs change, you can go back to the Item list and change your selling prices. But if you handle a wide variety of work and want to be ready to estimate almost anything, or if you use many unique items grouped together in a couple of dozen phases to make your estimates, you'll be asking too much from QuickBooks Pro.

In this chapter, we'll just show you how to enter a summary estimate into QuickBooks Pro.

Using a Summary Estimate You Make Outside of QuickBooks Pro

Before you transfer a summary estimate, it's a good idea to set up an item in QuickBooks Pro for each phase you use in your estimating program. See Chapter 4 for information on how to set up items. For example, suppose you decide to use our job phases partially shown in Figure 12-1. Then in QuickBooks Pro, you should set up these items. If you used our *company.qbw* data file to set up your own company data file, these items are already set up for you.

You can also create subitems to get more detail on the phases. For example, you could set up an item for the Plans and Permits phase with subitems for plans, building permits, and city license. Subcontractors may need an entirely different list of items — one that divides their work into appropriate categories.

Once you've entered the items (or modified existing items), you can manually transfer the information from each phase in a summary estimate to a matching QuickBooks Pro item. Let's walk through entering an estimate in QuickBooks Pro.

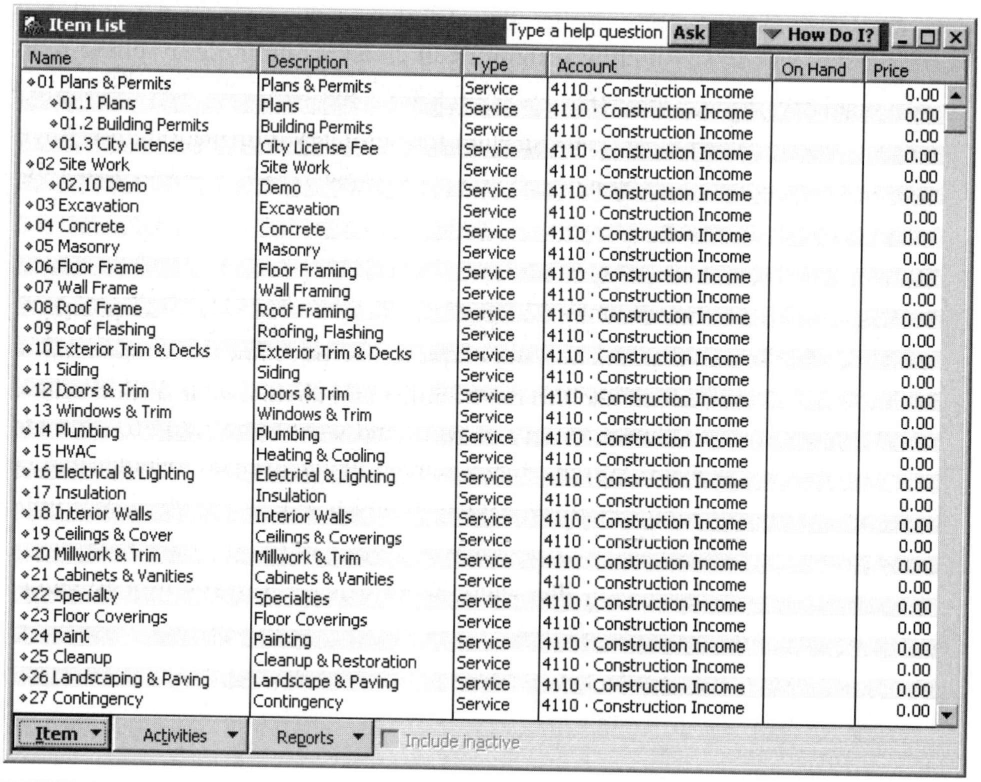

Name	Description	Type	Account	On Hand	Price
◆01 Plans & Permits	Plans & Permits	Service	4110 · Construction Income		0.00
◆01.1 Plans	Plans	Service	4110 · Construction Income		0.00
◆01.2 Building Permits	Building Permits	Service	4110 · Construction Income		0.00
◆01.3 City License	City License Fee	Service	4110 · Construction Income		0.00
◆02 Site Work	Site Work	Service	4110 · Construction Income		0.00
◆02.10 Demo	Demo	Service	4110 · Construction Income		0.00
◆03 Excavation	Excavation	Service	4110 · Construction Income		0.00
◆04 Concrete	Concrete	Service	4110 · Construction Income		0.00
◆05 Masonry	Masonry	Service	4110 · Construction Income		0.00
◆06 Floor Frame	Floor Framing	Service	4110 · Construction Income		0.00
◆07 Wall Frame	Wall Framing	Service	4110 · Construction Income		0.00
◆08 Roof Frame	Roof Framing	Service	4110 · Construction Income		0.00
◆09 Roof Flashing	Roofing, Flashing	Service	4110 · Construction Income		0.00
◆10 Exterior Trim & Decks	Exterior Trim & Decks	Service	4110 · Construction Income		0.00
◆11 Siding	Siding	Service	4110 · Construction Income		0.00
◆12 Doors & Trim	Doors & Trim	Service	4110 · Construction Income		0.00
◆13 Windows & Trim	Windows & Trim	Service	4110 · Construction Income		0.00
◆14 Plumbing	Plumbing	Service	4110 · Construction Income		0.00
◆15 HVAC	Heating & Cooling	Service	4110 · Construction Income		0.00
◆16 Electrical & Lighting	Electrical & Lighting	Service	4110 · Construction Income		0.00
◆17 Insulation	Insulation	Service	4110 · Construction Income		0.00
◆18 Interior Walls	Interior Walls	Service	4110 · Construction Income		0.00
◆19 Ceilings & Cover	Ceilings & Coverings	Service	4110 · Construction Income		0.00
◆20 Millwork & Trim	Millwork & Trim	Service	4110 · Construction Income		0.00
◆21 Cabinets & Vanities	Cabinets & Vanities	Service	4110 · Construction Income		0.00
◆22 Specialty	Specialties	Service	4110 · Construction Income		0.00
◆23 Floor Coverings	Floor Coverings	Service	4110 · Construction Income		0.00
◆24 Paint	Painting	Service	4110 · Construction Income		0.00
◆25 Cleanup	Cleanup & Restoration	Service	4110 · Construction Income		0.00
◆26 Landscaping & Paving	Landscape & Paving	Service	4110 · Construction Income		0.00
◆27 Contingency	Contingency	Service	4110 · Construction Income		0.00

Figure 12-1
The Item list in *sample.qbw*.

To create a new estimate in QuickBooks Pro using summary figures from an outside estimating program:

- From the **Customers** menu, choose **Create Estimates**.
- From the drop-down **Customer:Job** list, select the customer. In our example shown in Figure 12-2, we selected Pearl Downey:House-New Construction.
- Enter **Date** and **Estimate No**.
- To enter each line item for the estimate, click in the **Item** column, and select the item that matches the item in the outside summary estimate from the pull-down list. The description of the item will fill in automatically if you entered it when you created the item.
- In **Qty** and **Cost**, enter the figures for the item from the outside summary estimate. Amount will be calculated for you by multiplying Qty by Cost. The estimate form in Figure 12-2 has been customized to show columns for Qty, Estimate, Total, Markup, and Revenue, instead of Qty, Cost, Amount, Markup, and Total. We'll step you through customizing an estimate form in the next section.

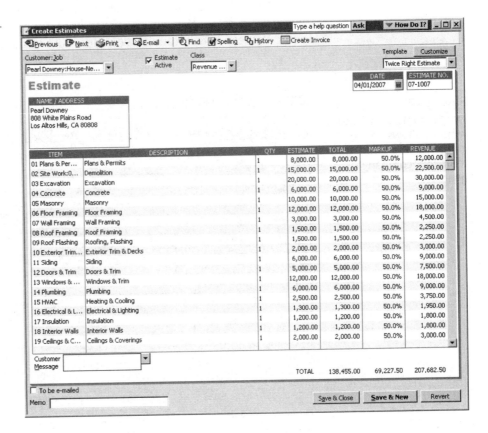

Figure 12-2
Use the Create Estimates window to quickly and easily create a new estimate for a customer.

Chapter 12: Estimating

- In **Markup**, enter the number to add a lump sum markup. To add a percentage, enter the number and a percentage sign (say 33%). QuickBooks Pro will calculate the appropriate markup, add it to the Amount, and show the result in Total.

- Click **Save & Close** to create the estimate.

Customizing an Estimate Form

You can change the way an estimate looks both on paper and on your computer screen. To do this:

- From the **Customers** menu, choose **Create Estimates**.
- Click the **Customize** button.
- In the Select an Action window, click **New**.
- In the Customize Estimate window, enter a new estimate template name. In our example shown in Figure 12-3, we entered **Twice Right Estimate.**

Take a look at each of the seven tabs in the Customize Estimate window. You can make changes to the form, both in what appears on the screen and what's printed. You can change titles in the header, fields, columns, and footer. You can add or remove columns, both in the printed estimate and on

Figure 12-3
The Customize Estimate window allows you to customize the estimate to fit your company's needs.

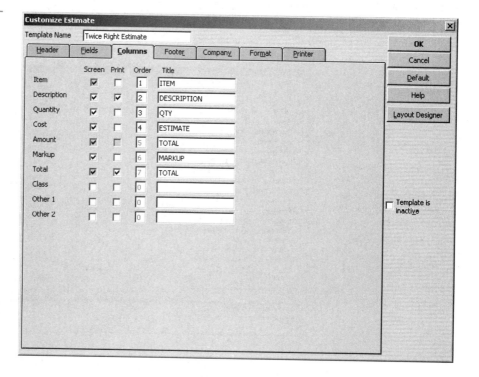

146 *Contractor's Guide to QuickBooks Pro*

the screen. You can even add your company logo or any picture bitmap (bmp file) at the top of the estimate using the Use Logo box on the Company tab. Use the Layout Designer to change estimate column widths.

▌ When you're done, click **OK** to save your new estimate template.

Memorizing an Estimate

It's a good idea to memorize an estimate that has line items you use frequently. For example, you could create an estimate for a kitchen addition, and memorize it as Basic, Medium, or High-end Kitchen. Then the next time you have a kitchen addition job to estimate, simply open the memorized estimate that's most like your new job and make changes to the items that are different.

Once you memorize an estimate, you can use it as a template — like a cookie cutter to stamp out more estimates with the same items, but with different customer or job names. This can save you many hours of estimating time. There's no need to start from scratch every time you estimate a job.

To memorize an estimate:

▌ With the estimate you want to memorize on your screen, from the **Edit** menu, choose **Memorize Estimate** and click **OK**.

▌ In the Memorize Transaction window, enter a name for the estimate. In our example shown in Figure 12-4, we entered *Kitchen Remodel*.

▌ Click **Don't Remind Me** because there's no need to be reminded regularly of a memorized estimate.

▌ Click **OK** to memorize the estimate.

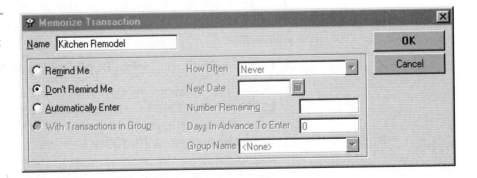

Figure 12-4
Memorize estimates that contain line items you use frequently. Be sure to give the memorized estimate a distinctive name, such as Kitchen Remodel, so it's easy to find when you need it.

To use a memorized estimate:

- From the **Lists** menu, choose **Memorized Transaction List**.
- In the Memorized Transaction List window, select the memorized estimate you want to use.
- Click **Enter Transaction**.

Now you'll see a Create Estimate window with nothing in Customer:Job and Name/Address.

- From the pull-down Customer:Job list, select the customer. Or select **Add New** to add a new customer. For more information on adding customers, see Chapter 7, Customers.

Of course, you'll probably never have two identical estimates for two customers. Usually, you'll have to change at least a few things in a memorized estimate. You'll probably need to change the quantities and costs, and add or delete lines.

To add (or delete) a line to an estimate:

- Click in the **Item** column.
- From the **Edit** menu, choose **Insert Line** (or **Delete Line**) and make the necessary changes.
- Click **Save & Close** and you've got a new version of an old estimate. Note that the memorized estimate hasn't changed at all. You can still use it for another estimate.

QuickBooks Pro lets you enter multiple estimates for one job. The advantage of having multiple estimates is that you can easily track the changes to the original estimate and proposed contract amount. Then, as the scope of work changes, you can compare the original estimate to the existing estimate and review it with the customer.

To enter a revised estimate, you should first create an estimate as described in the beginning of this chapter. Then, with the original estimate on your screen:

- From the **Edit** menu, select **Duplicate Estimate**.
- When the estimate is duplicated, make sure you change the **ESTIMATE NO.** field to reflect the revision number. For example, if your original estimate number was 01-1001, your revised estimate should be entered as 01-1001R1. If you have another revision, you should duplicate the latest estimate (01-1001R1 in our example) and enter R2 (or 01-1001R2) in the **ESTIMATE NO.** field.
- Click **Save & Close** to create the revised estimate.

Estimates and Progress Billing

After you create an estimate, you can easily create an invoice based on the estimate, using a progress billing invoice. For information on how to create a progress invoice, see Chapter 13. Before you create the invoice, you can see a summary of what you've billed to date, the current charges, and what will be billed later — all by phase of construction (or the items you've set up). So your final invoice is sure to cover everything in the job you haven't billed previously. You can make changes in the estimate any time you want.

From an estimate, you can:

1. Bill 100% — everything that remains to be billed.
2. Bill certain items only. You select which phases you want to bill.
3. Designate a percentage to be billed on any item you select.

If you handle larger, complex jobs that require multiple payments, progress billing may be a godsend. But progress billing only works on estimates, not invoices. To do progress billing you have to start with an estimate.

That's about all there is to know about bringing summary estimates into QuickBooks Pro.

Getting Detailed Estimates

There are ways you can use QuickBooks Pro to produce detailed estimates. One way involves creating a separate Items list in QuickBooks Pro — a somewhat complex task that requires more knowledge of QuickBooks Pro than we cover in this book. We wrote this book to take you step-by-step through a simple way to set up your bookkeeping in QuickBooks Pro. If you're comfortable with QuickBooks Pro and our basic setup, and are willing to give it a try, look in Appendix A. You'll find full instructions on setting up and using detailed cost estimates.

A far better option is Craftsman's Job Cost Wizard. Using this software, you can make estimates using Craftsman's estimating program, National Estimator, and import them directly into QuickBooks Pro. From there, you can track job costs for each item or category in an estimate, compare actual with estimated costs for each part of any job, and create and send invoices. Job Cost Wizard, National Estimator, and a 200-page cost estimating database are included on the CD in the back of this book. Appendix B gives detailed step-by-step instructions on how to use the National Estimator program to estimate and track job costs.

Chapter 13

Receivables

In this chapter, and the next, we'll show you how to use QuickBooks Pro on an accrual basis. The accrual basis of accounting means that you track how much money is owed to you (Accounts Receivable) and how much money you owe (Accounts Payable).

The accrual method has both benefits and drawbacks. Some of its benefits are:

- you can easily find out how much money you billed to customers vs. how much it cost you to earn that income
- you can easily find out how profitable your business was during a particular time period
- your job cost reports will include all costs and invoices to date

The major drawback to the accrual method is that it takes time to enter the transactions (bills and invoices). You have to enter every transaction as you receive or generate it. This requires vigilance to keep up to date. If you're a small contractor with little or no help, or no time to do bookkeeping, you're probably operating your business on a cash basis. If so, you can skip to Chapter 15, or keep reading to prepare for the time when you might switch to the accrual method.

Now on to the subject of this chapter — managing your receivables using the accrual system of accounting. Of course, you won't have any receivables if you don't send out bills, and contractors don't like to send out bills. However, with QuickBooks Pro it's so simple there's no excuse for having a backlog of invoices that you need to send out. When you've completed a job, or at designated stages during the job, you'll want to prepare invoices to submit to your customer for payment.

In this chapter we'll show you:

- four different ways to invoice a customer
- how to handle change orders

- how to handle retainage
- how to record customer payments
- how to enter a deposit
- how to record job deposits (customer advances)

Four Ways to Invoice a Customer

The way you prepare an invoice will depend on the terms of your contract. The main types of contracts are:

- set price (usually a small and simple verbal contract)
- time and materials or cost plus
- fixed price (usually a contract requiring a certain amount be paid when a certain percentage of the work has been completed)
- progress or AIAG702 structure

Creating a Set Price Invoice

Sometimes you need to get an invoice out the door quickly. This is usually for non-time-and-materials billings — perhaps a small, short-term job that doesn't require an estimate. For example, suppose a customer phones you to come out and fix a hole in a wall and you tell her you'll come and fix it today. Since she's a good customer who frequently refers your company to others, you agree to do the job for a flat $250. In this case, there's no estimate and the invoice is billed based on a set price.

Here's how you would enter a set price invoice:

- From the **Customers** menu, choose **Create Invoices**.
- From the pull-down **Customer:Job** list, select the customer. The Bill To section will fill in automatically if you completed that part of the customer record. See our example in Figure 13-1.
- Enter **P.O. No.** and any other general information you need.
- From the pull-down **Class** list, select **Revenue - Job Related**.
- Click in the **Item** column and select an item from the drop-down list. Keep in mind that your items represent job phases. **Description** will fill in automatically if you entered them when you set up the item. Change description to fit the job. For this example, we used item 18 Interior Wall with a description of Fix hole in wall of guest bedroom, and a rate of $250.00.

Figure 13-1
In the Create Invoices window, enter each item you want to appear on a printed invoice. Make sure the Customer:Job and Bill To address are correct.

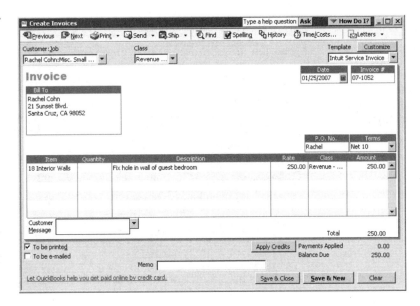

- Enter **Quantity** and **Rate** if necessary. Amount will be calculated automatically.

- From the pull-down **Terms** list, select the appropriate term. We've used Net 10 in our example.

If a term you normally use isn't on the drop-down list, you can add it. There are two types of terms — a standard term, which requires payment within a certain number of days, or a date driven term, which requires payment by a certain date. To add a standard term:

- From the pull-down Terms list, choose **Add New**. See our example in Figure 13-2.

- In the New Terms window, in **Terms**, enter a name for the new term.

- Click **Standard**.

- In **Net due in**, enter the number of days the customer has to pay the invoice.

- In **Discount percentage is** and **Discount if paid within**, enter a discount percentage and number of days required for the early payment discount if you offer such a discount.

- Click **OK** to create the new term.

Figure 13-2
Use the New Terms window to define the terms on your invoices. Use a Standard term to require payment within a number of days.

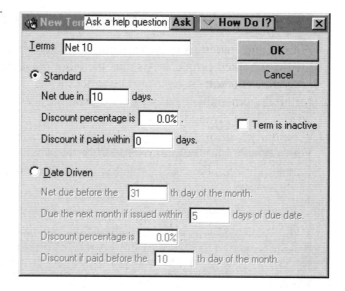

You set up a date driven term the same way, except you click Date Driven and enter the day of the month that the invoice is due. In our example, we want all invoices paid by the last day of the month, so we entered 31, as shown in Figure 13-3. You can also push back the due date if the invoice is issued late in the month, so your customers have plenty of time to pay their bills on time.

Now let's get back to the invoice. After you enter all the items you need:

▌ In the Create Invoices window, click the **Print** icon to print the invoice.

When you're satisfied the invoice is correct:

▌ Click **Save & Close** to close the invoice window. The invoice will automatically be added to Accounts Receivable.

Figure 13-3
Use a Date Driven term to require payment by a certain date.

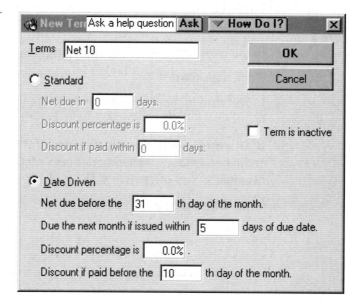

Creating a Time-and-Materials or Cost-Plus Invoice

A time-and-materials contract means that you agree to complete a job for the documented project costs plus the labor costs (at certain rates) for your crew and yourself. A cost-plus contract means that you bill your customer for all documented costs, plus a fee (generally a percent of the documented costs). These arrangements have the least risk and the least reward. You're essentially working for wages. It's very important that you make sure all your expenses are being covered with these types of contracts. Any costs that slip through the cracks and don't get billed are a complete loss that you can't recover.

Before you actually create a time-and-materials invoice for a customer, you must enter the job costs you incurred. You must also enter timecards for the employees who worked on the job. Information on how to enter job-related costs is in Chapter 14. For information on how to enter payroll, see Chapter 15.

In this example, we'll show you how to enter a time-and-materials invoice for a job. At this point you should have already entered bills or checks to vendors and assigned the costs to the job as well as payroll, if you have employees. The example screen shots are taken from the *sample.qbw* file. If you aren't prepared to enter your own time-and-materials invoice, you can open *sample.qbw* and follow along with our example. However, as you enter data into the sample data file, the data in the file will change. Consequently, the windows you see may be somewhat different from the examples we show in this chapter.

To create a new time-and-materials invoice:

- From the **Customers** menu, choose **Create Invoices**.

- From the pull-down **Customer:Job** list, select the customer. In our example, we use the Jose & Marie Sanchez: Cottage - New Construction job. If you made an estimate before preparing the invoice, you'll see a dialog box that tells you there's an estimate for the job and asks you if you want to create the invoice based on that estimate. If you get this dialog box, click **Cancel**.

- At the top of the Create Invoices window, click **Time/Costs** to get to the Choose Billable Time and Costs window.

- In the Choose Billable Time and Costs window, click the **Time** tab if you're charging for payroll costs and using time tracking. Make a checkmark in the Use column to select the billable time. See Figure 13-4.

- In the Choose Billable Time and Costs window, click the **Items** tab. Make a checkmark in the Use column to select the billable items. In our example, shown in Figure 13-5, we checked Roof Framing, Excavation materials, Exterior Trim & Decks, and Siding.

Figure 13-4
The Time tab displays a list of all time charged to the job by employees.

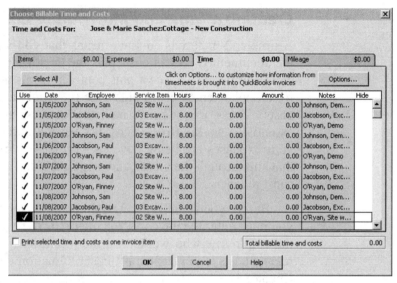

Figure 13-5
Use the Items, Expenses, and Time tabs to create the line items of a time-and-materials invoice.

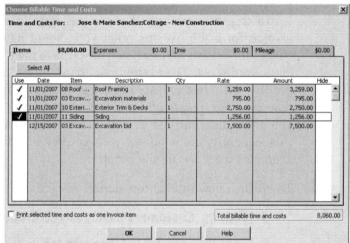

- If you want the selected time and costs to appear as one item on the invoice, click **Print selected time and costs as one invoice item**.
- Click **OK**.
- Back in the Create Invoices window, enter the rate you charge for the employee in the **Rate** column. In our example, we charge $60 per hour for Paul Jacobson's labor. Amount is calculated from the rate.
- Click **Print** to print the invoice.

When you're satisfied that the invoice is correct:

- Click **Save & Close** to close the invoice window. The invoice will automatically be added to Accounts Receivable.

Your completed time and materials invoice will be similar to the example we show in Figure 13-6.

Figure 13-6
A completed time-and-materials invoice.

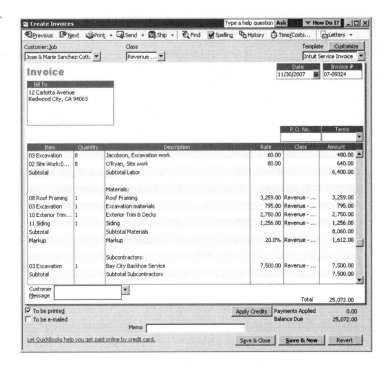

Creating a Fixed Price Contract Invoice

A fixed price (also called lump sum) contract means that, except for change orders, you agree to complete the job as specified in the plans and contract documents for an agreed amount. You assume the entire risk if the job goes over budget. But you also have the opportunity to exceed your estimated profit if you can bring the job in under the contract amount.

When you use a fixed price contract, you bill a specified amount of the contract price when you complete designated stages of the job. The specified amount is typically less than the total contract price. As an example, you might have a billing schedule like this one:

Billing #1: 40% of contract payable on completion of rough framing

Billing #2: 20% of contract payable on completion of drywall

Billing #3: 30% of contract payable on completion of final flooring

Billing #4: 10% of contract payable on completion of signing punch list

The customer or the lender usually makes an inspection to authorize a billing against the contract.

To make an invoice for a fixed price contract, you need to specify the percentage of completed work due on the invoice. And you must create an item so you can enter this percentage. Here's how to create the item:

■ From the **Lists** menu, choose **Item List**. In the Item List window, pull down the **Item** menu and choose **New**.

- For **Type**, select **Service**.
- In **Item Name/Number**, enter **Fixed Price Billing**. See Figure 13-7.
- In **Description**, enter a generic description. This description will automatically fill in on the invoice when you use the item. In the example shown in Figure 13-7, we used "_% due upon completion of __." You can fill in the blanks in this description when you complete the invoice the item appears on.
- In **Account**, select your Construction Income account. In our example, we used 4110 Construction Income.
- Click **OK**.

To create a fixed price invoice you can follow our example in Figure 13-8:

- From the **Customers** menu, choose **Create Invoices**.
- From the **Customer:Job** drop-down list, select a customer. For our example, we've used Dennis and Judy Bales: Addition. If you made an estimate before preparing the invoice, you'll see a dialog box that tells you there's an estimate for the job and asks if you want to create the invoice based on that estimate. Click **Cancel**.
- From the pull-down **Class** list, select **Revenue - Job Related**.
- Click in **Item** and from the pull-down **Item** list, select **Fixed Price Billing**.
- In **Quantity**, enter as a decimal the percentage of the contract you're currently billing.
- In **Description**, enter whatever text you need for the invoice.
- In **Rate**, enter the full contract amount. Amount will be computed for you by multiplying the percentage you entered in Quantity times the full contract amount.

Figure 13-7
Use this item when you create an invoice for a fixed price contract.

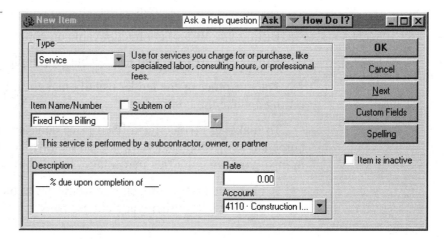

Figure 13-8
A completed fixed price contract invoice.

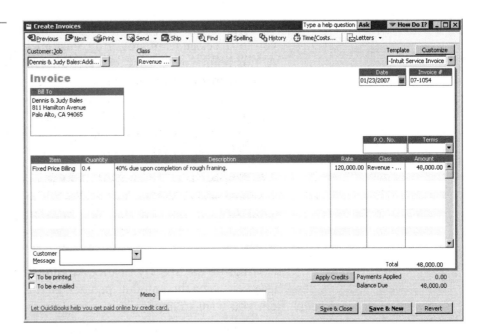

■ Click **Print**. Check the printed invoice carefully to make sure it's correct. Then print two copies so you have one for your customer file.

■ Click **Save & Close**.

Creating a Progress Invoice

Before you can create a progress invoice for a job, you have to do two things:

■ Create an estimate for the job using items. See Chapter 12 if you need information on creating estimates.

■ Make sure the Progress Invoicing preference is active.

To turn on the progress invoice preference:

■ From the **Edit** menu, choose **Preferences** and click the **Jobs & Estimates** icon and the **Company Preferences** tab.

■ Click **Yes** to answer the question **Do You Do Progress Invoicing?**

■ Click **OK**.

Now, to create a progress invoice:

■ From the **Customers** menu, choose **Create Invoices**.

■ From the drop-down **Customer:Job** list, select the customer. For our example, we've used Pearl Downey: House - New Construction.

- If you're following our example, you'll see a dialog box that allows you to select an estimate for the customer, as shown in Figure 13-9. At this point you'll need to select the estimate you want to create a progress invoice against.

- Select **OK**.

- This takes us to the Create Progress Invoice Based On Estimate window shown in Figure 13-10. Now we have to decide how you want to use the estimate to create the invoice. Make your selection here based on the type of contract you have with the customer. For our example, we select the last option — Create invoice for selected items or for different percentages of each item.

- Click **OK**.

This brings up the Specify Invoice Amounts for Items on Estimate window. Here you enter the amount, percentage, quantity, and/or rate due for each item you want to appear on the invoice. To specify invoice amounts from an estimate:

- In the **Curr %** column for the item, enter the new percentage. Amount will fill in based on the new percentage. In our example shown in Figure 13-11, we invoiced 100% of Plans & Permits, Site Work, Excavation and Concrete.

- Or, in the **Amount** column for the item, enter the new amount. Curr% will fill in based on the new amount.

- In the **Quantity** column for the item, enter the new quantity.

- In the **Rate** column for the item, edit to enter the new rate.

- Click **OK** to get back to the Create Invoices window.

- Back in the Create Invoices window, click **Print** to print the invoice, shown in Figure 13-12.

Check your printed invoice carefully. When you're satisfied that the invoice is correct:

- Click **Save & Close** to close the invoice window. The invoice will automatically be added to Accounts Receivable.

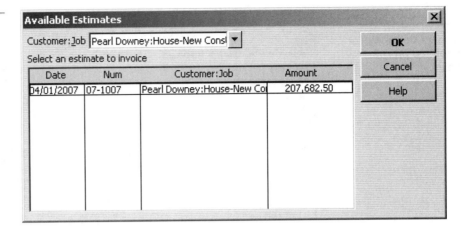

Figure 13-9
QuickBooks Pro lets you choose to invoice from an existing estimate or start the invoice from scratch.

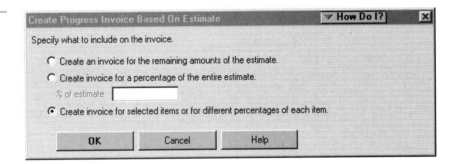

Figure 13-10
When you use an existing estimate to create an invoice, you have three choices for how to create the invoice.

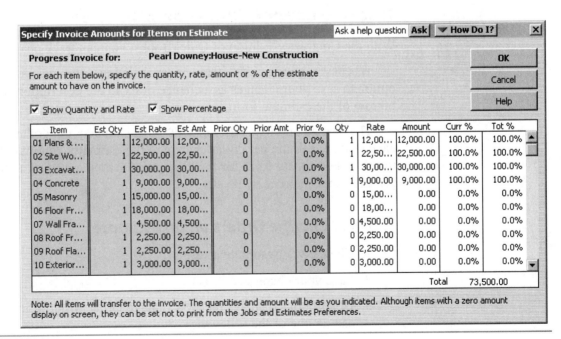

Figure 13-11
Use this window to choose which items from an estimate to include in an invoice.

Figure 13-12
This is an example of a completed progress invoice.

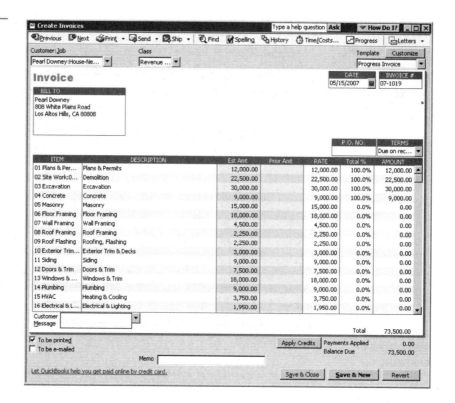

Tracking Change Orders on Estimates

With the Contractor Edition of QuickBooks, you can track change orders.

Whenever you modify items in an existing estimate, you have the option to save your changes as a change order in the description field at the bottom of the estimate form.

The change order specifies exactly what changed, the dollar amount of each change, and the net dollar change to the estimate. You may then wish to print the estimate form for the customer for final approval. (You cannot print the change order by itself.)

To Track Change Orders on an Estimate

1. Change your estimate as usual.

2. Save your changes.

3. QuickBooks displays the Add Change Order window, in which you can do one of the following:

■ Click **Add** to add the displayed text to the bottom of the estimate. See Figure 13-13.

If you don't like the way the change order appears, you can edit the text in the Add Change Order window before clicking **Add**, or you can edit the change order on the estimate form later.

■ Click **Do Not Add** if you want the estimate to be updated with your changes without explicitly listing the change order at the bottom of the form.

■ Click **Cancel** to return to the estimate form and reconsider your changes.

Tip: If you want to save your original estimate before creating a change order, right-click in the Create Estimates window and choose Duplicate Estimate. Either save and edit the duplicate estimate, or display and edit the original estimate. Note that if you edit the duplicate estimate before you save it, a change order won't be created. Change orders are generated only on existing estimates that have previously been saved.

Figure 13-13
To create a change order in the Contractor Edition, you simply change an existing estimate.

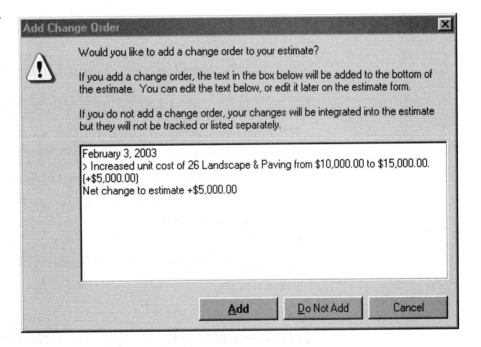

Reviewing and Changing an Estimate

1. Display the estimate. From the Customer menu, select Create Estimates. Hit **Previous** button until you find the estimate you want to modify.

2. (Optional.) Click **Print** to create a copy of the original estimate for your records.

3. Enter your changes on the estimate form. QuickBooks automatically recalculates totals.

4. Save your changes.

How to Handle Retainage

Sometimes you'll be required to hold back 10 percent of your total progress payments (billing only for 90 percent of the contract price) as a retainer until you've completed the project satisfactorily. After the job is complete and it's agreed that you can collect the retainage, you can bill the customer for the remaining 10 percent of the project cost. Here's how to handle this type of billing:

Step 1— Setting Up an Account for Retentions Receivable

First you need to set up an asset account for retainage. If you used our example Chart of Accounts, the asset account number for retainage is 1320 - Retentions Receivable. If you don't have an asset account for retentions receivable, you'll need to follow these steps to enter a new account:

- From the **Lists** menu, choose **Chart of Accounts**.
- In the lower part of the Chart of Accounts list window, click **Account** and choose **New**. See Figure 13-14 for an example of a New Account window.
- Fill in the information for the Retentions Receivable account.
- Click **OK**.

Step 2 — Setting Up an Item for Retentions Receivable

If you don't have an item for Retentions Receivable, follow these steps to create a new item:

- From the **Lists** menu, choose **Item List**.
- Click **Item** in the lower part of the Items List window. From the pull-down menu choose **New**. See Figure 13-15 for an example of a New Item window.

164 Contractor's Guide to QuickBooks Pro

- In **Type**, select **Other Charge**.
- Fill in the fields for the Retentions Receivable item.
- Click **OK** to create the new item.

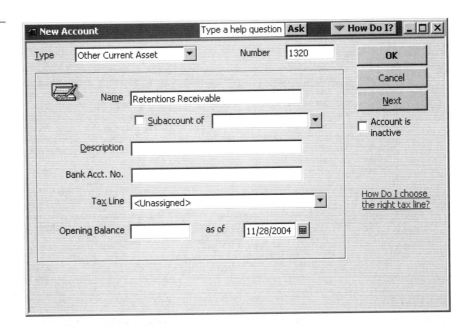

Figure 13-14
Use the New Account window to create an account for retentions receivable.

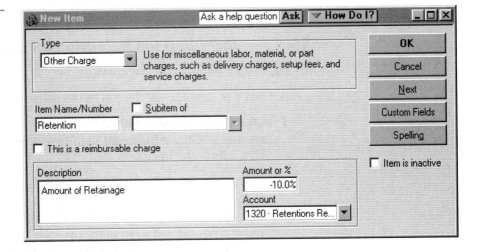

Figure 13-15
Use the New Item window to create an item for retention.

Chapter 13: Receivables

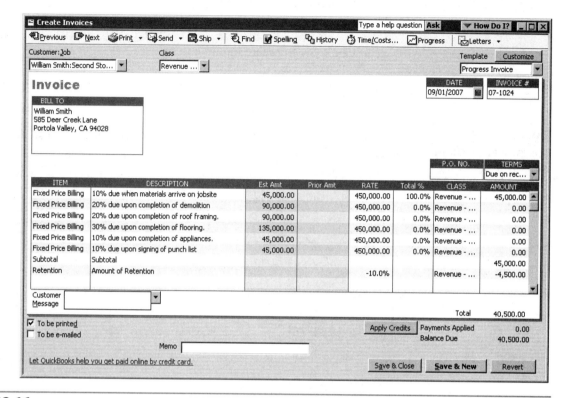

Figure 13-16
When you include a retention amount on an invoice, the balance due on the invoice is lowered by that amount.

Step 3 — Including the Retention on an Invoice

To include the retention on an invoice:

- Create an invoice for the job in the usual way.

- In the Create Invoices window, click on a blank line in the **Item** column and select the **Subtotal** item from the pull-down Item list. This will subtotal the invoice.

- Then select the **Retention** item from the pull-down Item list. The retention item will automatically deduct 10% from the subtotal line listed above. In Figure 13-16, the original amount of the invoice was $45,000. We entered a retainage of −$4,500. The final amount of the invoice ends up to be $45,000 − $4,500 = $40,500.

- Click **Save & Close**.

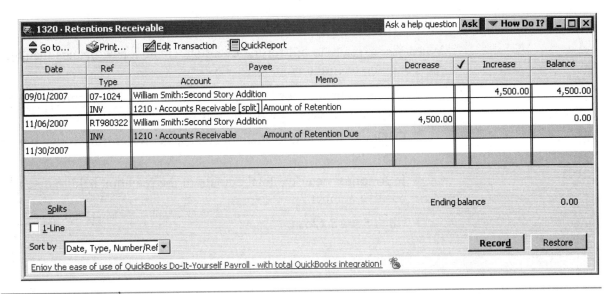

Figure 13-17
The Retentions Receivable register shows transactions in the Retentions Receivable account.

Recalling Who Owes You a Retention and How Much You're Owed

An easy way to determine the amount is to look at the register of your Retentions Receivable account. If you used our *company.qbw* data file to set up your company, you can use the memorized Retentions Receivable Report to see this information. You'll find more information on this report in Chapter 17, Reports. Otherwise, to see your retentions receivable:

- From **Lists** menu, choose **Chart of Accounts**.

- In the Chart of Accounts list window, select your **Retentions Receivable** account. In our example shown in Figure 13-17 it's 1320 - Retentions Receivable.

- From the bottom of the Chart of Accounts window, click **Activities** and choose **Use Register** from the pull-down menu. You should see a register that shows all the activity in your Retentions Receivable account. See Figure 13-17.

- Click the **X** in the upper right of the window to close the register.

Also, in Chapter 17, Figure 17-11, you'll learn how to print or display to screen the Retentions Receivable report. This report will summarize who owes you for retention and how much they owe.

Invoicing for Retention

When you're ready to bill for the retention, create a new invoice that contains only the retention item.

▮ Create an invoice for the job in the usual way. See Figure 13-18 for our example, using a Retention Invoice.

▮ Click in the **Item Code** Column and select **Retention** from the pull-down list.

▮ In **Amount**, enter the total amount of the retention held to date as a positive number.

▮ Click **Save & Close**.

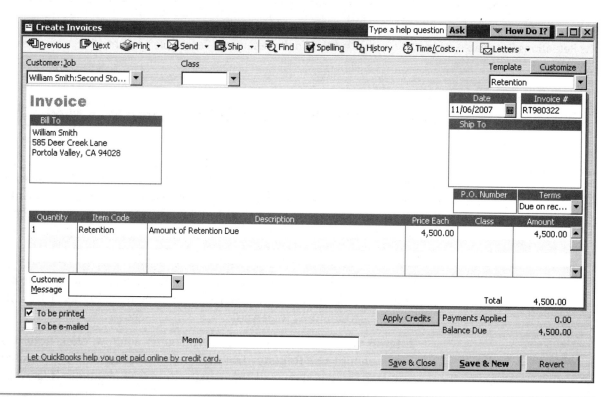

Figure 13-18
When you bill for retention, create a new invoice for just the retention.

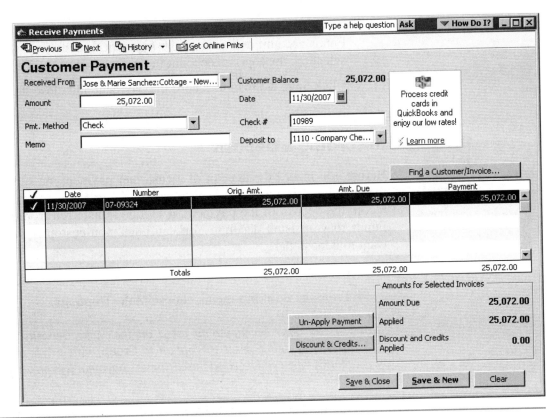

Figure 13-19
Use the Receive Payments window to record each payment you receive from a customer.

Recording a Payment You Receive

When you create and send an invoice to a customer, QuickBooks Pro creates a record in the Accounts Receivable register. When you receive a payment from the customer, you record receipt of the payment in the Receive Payments window.

To record a payment:

- From the **Customers** menu, choose **Receive Payments**.

- In the Receive Payments window, select the customer from the pull-down **Received From** list. In our example shown in Figure 13-19, we show receipt of a payment from customer Jose & Marie Sanchez.

- Fill in the date you received the customer's payment, the amount paid, the payment method, and the check number if the customer paid by check.

- If you want to hold the payment until you deposit the check, select your Undeposited Funds account in the **Deposit to** field. In our example, it's account 1390. We'll show you how to record the deposit in the next section. If you want to bypass the Undeposited Funds account and deposit the payment directly in a bank account, select the appropriate account in the **Deposit to** field.

▌ Click **Save & Close** to record the payment and close the window.

Recording a Deposit

Before you record the deposit of a customer payment, be sure you record the payment using the steps we showed in the previous section. When you record the deposit of the payment, it will be transferred from Undeposited Funds to your bank account. You can enter non-customer payments you want to deposit here.

To record a deposit:

▌ From the **Banking** menu, choose **Make Deposits**.

> *Note: If you don't see anything listed on the Make Deposits screen, you probably didn't select the option to* Group with other undeposited funds *when you received the customer's payment (see previous section).*

▌ In the Payments to Deposit window, click each customer check you're depositing. In our example shown in Figure 13-20, we select the payment from Jose & Marie Sanchez.

▌ Click **OK** to proceed to the Make Deposits window.

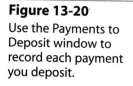

Figure 13-20
Use the Payments to Deposit window to record each payment you deposit.

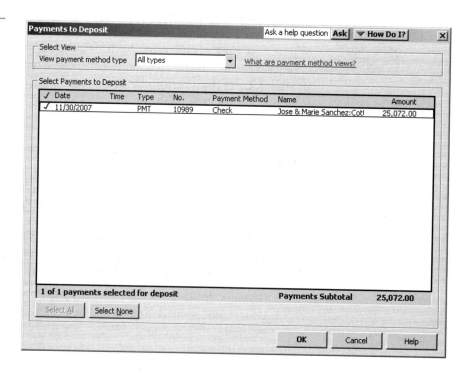

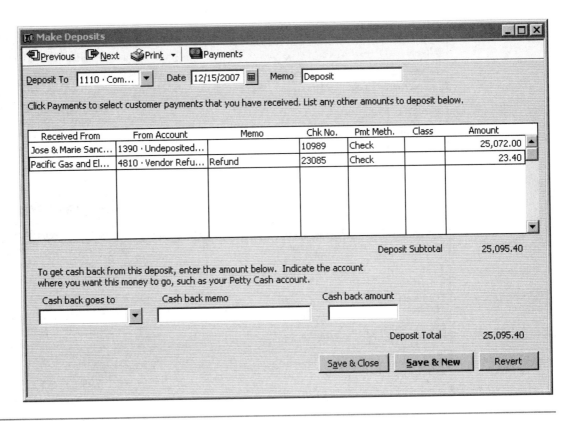

Figure 13-21
You can also use the Make Deposits window to record non-customer payments you deposit.

- In the Make Deposits window, add any non-customer payments that you'll be depositing. In our example shown in Figure 13-21, we deposit a refund from Pacific Gas & Electric for $23.40.
- Click **Save & Close** to record the deposit.

Recording a Job Deposit

A job deposit is money that belongs to the customer, but you hold it until you fulfill your contractual obligations. In other words, it's money in your hand, but you still need to earn it. Don't record it as income when you get it. Record it as a liability on your books even though you deposit the funds into your checking account. When the customer agrees you can apply the deposit to the job costs, or defaults on the contract, you can record the deposit as income.

Chapter 13: Receivables

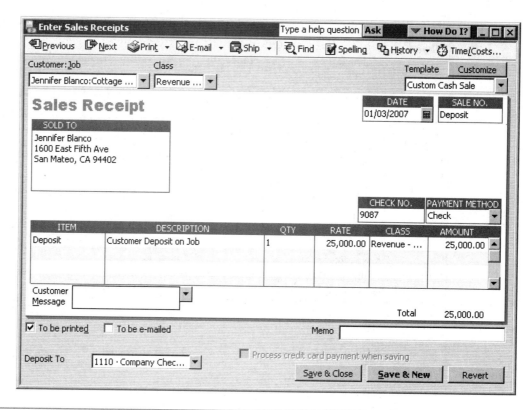

Figure 13-22
Use the Enter Sales Receipts window to record a job deposit as a liability.

To record a deposit as a liability:

- From the **Customers** menu, choose **Enter Sales Receipts**.

- From the pull-down **Customer:Job** list in the Enter Sales Receipts window, select the customer. In our example shown in Figure 13-22, we enter a customer deposit for $25,000 from Jennifer Blanco.

- In **Class**, select **Revenue - Job Related**.

- In **Date**, enter the date you deposited the funds.

- In **Sale No.**, enter **Deposit**.

- Select **Payment Method** from the drop-down list.

- In the **Item** column, select **Deposit** from the drop-down Item list. In **Quantity**, enter **1**. Enter **Rate**. Amount will be calculated.

- To print a message on the receipt, select a message from the **Customer Message** drop-down list. To add a new message, click **Add New** and create your own personalized message.

- Click **Print** to create the receipt for your customer.

172 *Contractor's Guide to QuickBooks Pro*

- At the very bottom of the Enter Sales Receipts window, click **Deposit To**. From the pull-down Account list, select the **Checking** or **Savings** account you want to put the funds in. Make sure you deposit this check by itself, not as part of a group deposit. When you reconcile the Checking or Savings account in QuickBooks Pro, you want this deposit to be listed alone.

- Click **Save & Close** to record the deposit.

To apply the deposit to an invoice:

- Follow the usual steps for creating an invoice for the customer.

- At the end of the invoice items, enter a deposit item using a negative rate. In our example shown in Figure 13-23, we subtracted $25,000 from the Jennifer Blanco:Cottage Progress invoice.

- Click **Save & Close** to create the invoice.

Now you know how to process invoices to customers in QuickBooks Pro.

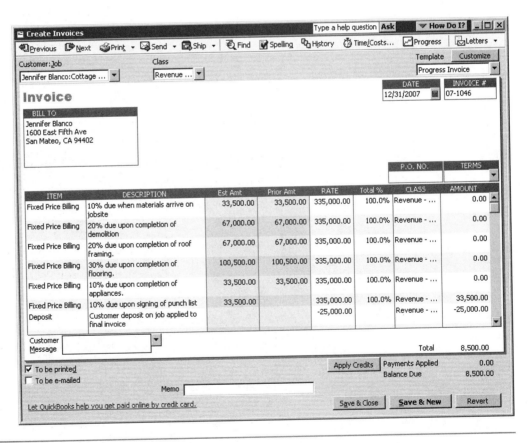

Figure 13-23
This Create Invoices window shows a deposit deducted from a final billing.

Chapter 14

Payables

We've based the information in this chapter on using QuickBooks Pro on an accrual method of accounting we discussed in Chapter 13. The advantages of the accrual method of accounting, as it relates to Accounts Payable, are:

- your job cost reports will include vendor bills that you haven't paid
- your Accounts Payable Report will help you track the vendors you owe money to
- your profitability reports will be more accurate

If you don't use an accrual method of accounting, you'll find the information you need on cash accounting in Chapter 16.

Here's what we're going to discuss in this chapter:

1. Entering purchase orders
2. Using purchase orders to track multiple draws and committed costs
3. Entering bills for job-related and overhead expenses without purchase orders
4. Selecting bills for payment
5. Printing checks
6. Payables and vendor workers' comp reports

To get job and non-job expenses into your accounting system accurately and completely, you'll have to enter who you're paying and what you're paying for. Be sure to enter a memo or note about each expense and put a record of the transaction in the proper account. Then your QuickBooks Pro reports will accurately reflect all your expenses.

You should already have set up your Chart of Accounts and other lists in QuickBooks Pro before going through this chapter. You should also open the *sample.qbw* file and use it to follow the examples in this chapter. Then you can try out the steps you need to go through without affecting any actual company information.

Creating and Using Purchase Orders

Purchase orders in QuickBooks Pro are a record of what you've ordered. They let you compare what you order with what you actually get. For example, suppose Twice Right Construction ordered cabinets from Arturo's Cabinets and created the purchase order shown in Figure 14-1. QuickBooks Pro will track the order date and the amount quoted for the cabinets. Then Twice Right doesn't have to rely on memory for this information.

So let's look at how to create a purchase order:

- From the **Vendors** menu, choose **Create Purchase Orders**.

If you don't see the command Create Purchase Orders in the Vendors menu, you need to turn on purchase order tracking. To do this:

- From the **Edit** menu, choose **Preferences**.
- Click **Purchases & Vendors**.
- Click the **Company Preferences** tab.
- Select **Inventory and purchase orders are active**.
- Click **OK**.

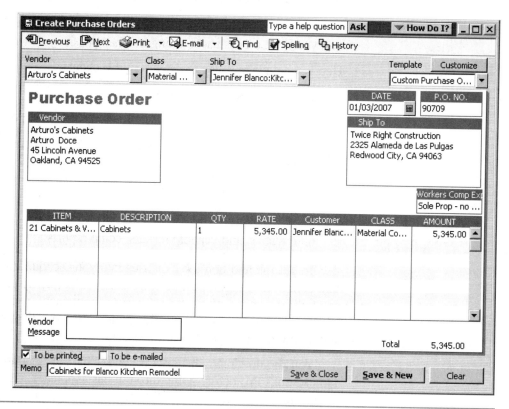

Figure 14-1
Use purchase orders to track items that you've ordered and check against what you receive.

Now back to our purchase order.

- In the top part of the form, enter the Vendor, the Ship To address (job site), the Class (materials), Date, PO number, and vendor's address.

- Click in the **Item** column and select the items you're ordering from the pull-down list. Change the Description and Amount fields as necessary.

- Click in the **Customer** column and select the customer you're ordering materials for from the pull-down Customer list.

- In **Memo** enter a brief description about this PO. This is important because the memo will appear when you move the PO into a Vendor bill.

- Click **Save & Close** or **Save & New** to enter another PO.

Using Purchase Orders to Track Multiple Draws and Committed Costs (Buyouts)

In QuickBooks Pro, you can assign each line item of a purchase order to a job. This feature will let you use the purchase order to track multiple payment draws and committed costs to subcontractors. Let's look at multiple draws first.

Multiple Draw Schedules

Suppose you want to track an agreement like the one we made with Ace Framing Contractors for a three-part draw. As soon as you make this type of agreement or contract, you should enter the draw schedule as a purchase order. In Figure 14-2 you'll see an example showing a purchase order where Twice Right Construction entered a three-part draw to Ace Framing for floor (Draw #1), wall (Draw #2), and roof (Draw #3) framing.

When Twice Right received Ace Framing's first bill for Draw #1 — Floor Framing, they used that same purchase order to record the payment of the first draw. Here are the steps:

- From the **Vendors** menu, choose **Receive Items and Enter Bill**.

- In the Enter Bills window, from the pull-down **Vendor** list, select the appropriate vendor. In our example it's Ace Framing.

- If you're following our example, you'll see the Open PO's Exist window, where you'll be warned that open purchase orders exist for the vendor. You're asked if you want to receive against one of them. Click **Yes** here to tell QuickBooks Pro that you want to use an existing PO.

Chapter 14: Payables

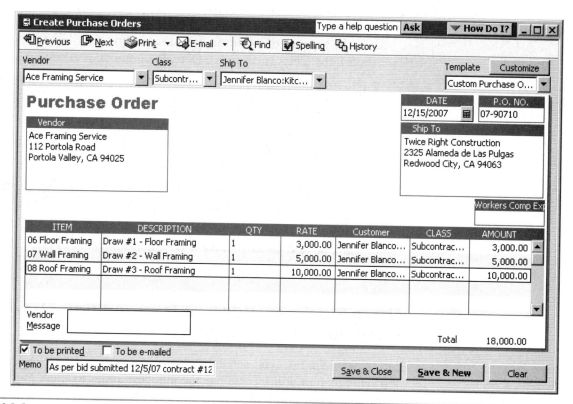

Figure 14-2
Use a purchase order to enter the parts of a multiple-payment draw schedule.

- In the Open Purchase Orders window, the vendor you selected appears along with a list of the purchase orders open for the vendor. Click in the ✓ column to select the appropriate open PO. In our example shown in Figure 14-3, we checked the PO dated 12/15/2007.
- Click **OK**.

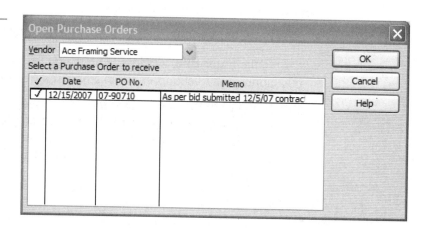

Figure 14-3
Use the Open Purchase Orders window to see a list of the open POs for a vendor.

178   *Contractor's Guide to QuickBooks Pro*

Back in the Enter Bills window, as shown in Figure 14-4:

- In **Date**, enter the date on the vendor's bill.
- In **Ref. No.**, enter the vendor's invoice number.
- In **Qty** enter 1 for any draw that you want to pay now. If you only want to pay half of the draw, enter 0.5 in Qty.
- In **Amount Due** enter the amount you plan to pay the vendor (or the draw). In our example, Twice Right will pay $3000, which is 100 percent of Draw #1 - Floor Framing.
- Click in the **Amount** column to enter the amount you're paying. In our example, it's $3,000 for Draw #1. The amount shown on the Items tab should be the same as the amount you're paying. The amounts for the other draws you're not paying will be 0.00.
- Click **Save & Close**.

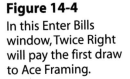
Figure 14-4
In this Enter Bills window, Twice Right will pay the first draw to Ace Framing.

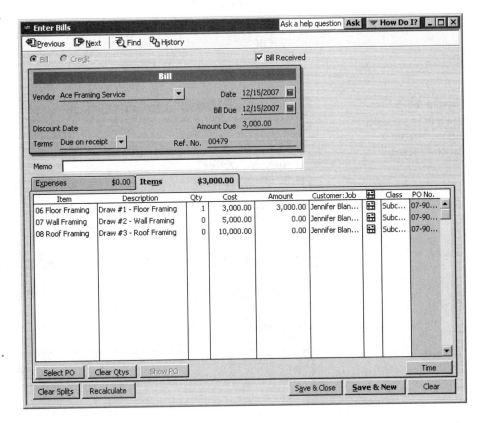

Committed Costs

You can also use a purchase order and the same method to track what you've committed to pay a subcontractor. For example, suppose you have a subcontractor who bid a project out at a certain price and you want to track the bid price to compare it to his final bill. To do this, you could set up a purchase order at the beginning of the project and compare it to the final bill you receive from the subcontractor. In our example in Figure 14-5, Twice Right used a PO to track the bid price (and/or contract price) from Blue Dolphin Plumbing for rough and finish plumbing. When Blue Dolphin Plumbing submits their bill, Twice Right can compare the original bid price to the original purchase order.

Tracking Unfilled Purchase Orders

QuickBooks Pro has a built-in report to track purchase orders that haven't been filled. You can run this report to list all subcontractors who haven't completed or submitted bills on a project. To generate this report:

- From the **Reports** menu, choose **Purchases** and then **Open Purchase Orders by Job**.

- In the Modify Report: Open Purchase Orders by Job window, click **OK**.

Figure 14-5
Use a PO to track a bid price from a subcontractor.

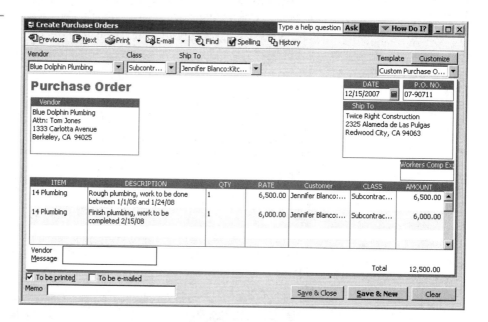

Important note! You may have a problem with this report if you make a partial payment to a subcontractor. The amount remaining after the partial payment will clear from the report unless you enter a number less than 1 in the Qty column in the Enter Bills window. To avoid this reporting problem, always enter a number *less than 1* when you release part of a PO. For example if you pay $1,500 of a $3,000 bid, enter 0.5 in Qty.

Entering Bills Without Purchase Orders

Entering Bills for Job-Related Expenses

Suppose you don't have the need (or time) to enter a purchase order before you get a vendor bill. In this case, you'll want to skip entering the purchase order and enter the bill directly. In Figure 14-6 we show an example of a bill Twice Right got from Eric Savage Electrician for $2,300 that they hadn't entered a purchase order for. To record the $2,300 they owe and get the fee into their job cost reports, they'll enter a bill in QuickBooks Pro. It's important to enter job-related costs on the Items tab when you enter a bill.

To enter a bill for an expense that's a job-related expense:

■ From the **Vendors** menu, choose **Enter Bills**.

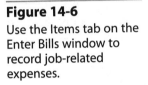

Figure 14-6
Use the Items tab on the Enter Bills window to record job-related expenses.

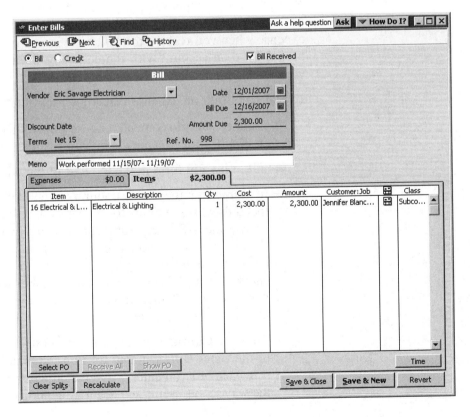

- Fill in the **Bill** form in the top half of the window.

- In **Memo**, enter a memo that tells what the bill is for. This memo will appear on billing reports such as A/P Aging Summary and Vendor Balance Detail. If you pay the bill by check and you *didn't* enter an account number when you set up the vendor's account, QuickBooks Pro puts the memo in the Memo field on the check. If you *did* enter an account number when you set up the vendor's account, QuickBooks Pro puts the account number in the Memo field of the check.

- Click the **Items** tab. Be sure that you enter job-related expenses on the Items tab of the Enter Bills window.

- Click in the **Item** column and select the items for the bill from the pull-down list.

- Click in the **Customer:Job** column and select the customer from the pull-down list.

- Click in the **Class** column and select the class from the pull-down list.

- Edit **Description** and **Amount** if necessary.

- Click **Save & Close** to record the bill.

Entering Bills for Overhead Expenses

Non-job related expenses are overhead expenses for things such as telephone, electricity, office supplies, etc. You should enter a bill for any expense when you receive the bill, even if you charged it to a charge card or you have an account with the company. Then you'll have the costs in your books as soon as it's been charged.

To enter a bill for an expense that's not a job-related expense:

- From the **Vendors** menu, choose **Enter Bills**. In our example shown in Figure 14-7, we show a bill from a telephone company.

- Fill in the **Bill** form in the top half of the window.

- In **Memo** (not the Memo column on the Expenses tab), enter a memo that tells what the bill is for. This memo will appear on billing reports such as the A/P Aging Summary. If you pay the bill by check and you *didn't* enter an account number when you set up the vendor's account, QuickBooks Pro puts the memo in the Memo field on the check. If you *did* enter an account number when you set up the vendor's account, QuickBooks Pro puts the account number in the Memo field of the check.

- Click the **Expenses** tab. It's important that you use the Expenses tab of the Enter Bills window for non-job expenses. You have to make sure each bill you enter is associated with the proper account. That's the way to get useful and accurate reports on your expenses.

Figure 14-7
Use the Expenses tab on the Enter Bills window to record overhead expenses. Make sure you enter the correct account.

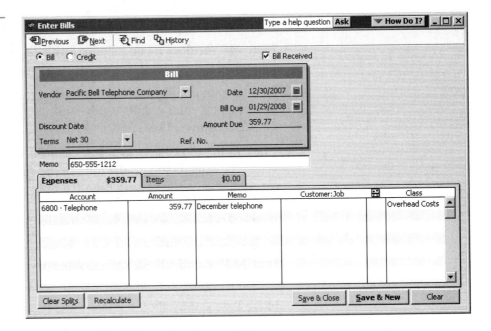

- Click in **Account** and select the appropriate expense account or accounts to assign the bill from the pull-down list. If you're paying a bill with multiple items and you need to split the amount over more than one expense account, change the amount that's automatically entered in the first line to match the line item in the bill. Then tab to the second line and make the entry for the next line item. Continue this way for each line item.

- In the **Memo** column, enter a description or account number (you won't have a customer:job for non-job related expenses).

- Click **Save & Close** to record the bill.

Selecting Bills for Payment

The Pay Bills feature in QuickBooks Pro lets you select the bills you want to pay. This is usually called a check run. In the construction industry, check runs usually happen every Friday or on the last day of the current month and the 15th day of the following month. In this section, we'll explain how to select the bills you want to pay. In the next section, we'll explain how to print the checks, if you have check stock.

To select bills for payment:

- From the **Vendors** menu, choose **Pay Bills**. QuickBooks Pro will list all unpaid bills or all bills due for the date range you enter.

- At the top of the Pay Bills window, next to **Show bills,** select **Due on or before** to enter a date to cover all the bills that you want to pay. Or select **Show all bills**.

- Click in the ✓ column to select each bill you want to pay. As you select the bills, compare the total in the **Amt To Pay** column against the total shown for **Ending Balance** in the Payment Account section of the Pay Bills window. This tells you if you have enough money in your checking account to cover all the bills you've selected to pay.

- To pay only part of a particular bill, change the amount shown in the **Amt To Pay** column. In our example shown in Figure 14-8, Twice Right paid $2,000 of the $3,000 owed to Ace Framing Service.

- In the Payment Method section of the Pay Bills window select **To be printed**. From the pull-down Payment Method list, select **Check**.

- To record the payments, click **Pay & Close**.

Printing Checks

To print your checks:

- From the **File** menu, choose **Print Forms** and then **Checks**.

- Put a check mark in front of each check you want to print. In our example shown in Figure 14-9, we select a $2,000 check to Ace Framing Service.

- Click **OK** to print the checks.

Be sure you have blank check stock in your printer, and that the number in the First Check Number field matches the first number on your checks in the printer.

Chapter 14: Payables

Figure 14-8
Select the bills you want to pay in the Pay Bills window.

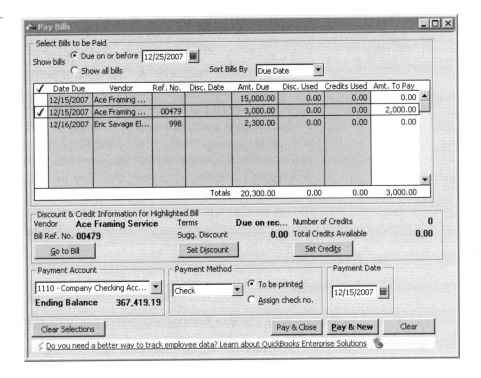

Figure 14-9
Select the checks you want to print in the Select Checks to Print window.

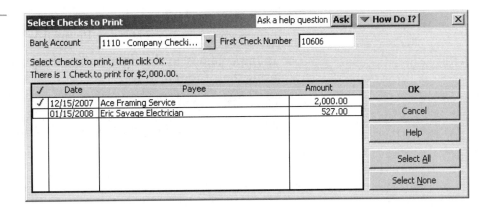

Contractor's Guide to QuickBooks Pro

Vendor Workers' Comp Reports

Staying on top of the workers' comp expiration dates for your subcontractors is a constant headache. As a matter of policy, you shouldn't even issue a purchase order, let alone pay a bill, for a subcontractor who hasn't supplied you with a current certificate of insurance for workers' comp. A workers' comp audit can cost you dearly if you can't show that your subcontractors had coverage at the time they provided their services. You can easily create a custom field in your vendors list to hold the date each subcontractor's workers' comp expires. You may find it helpful to get a report of your vendors and their workers' comp expiration dates. The *sample.qbw* file on the CD with this book has the custom field Workers' Comp Expires created for vendors. You'll find more information on the Vendor Comp Expirations report in Chapter 17, Reports.

Here's how to create a custom field:

- From the **Lists** menu, choose **Vendor List**.
- In the **Vendor List** window, click **Vendor** and then **Edit Vendor**.
- Click the **Additional Info** tab. The example in Figure 14-10 shows the Workers' Comp Expires custom field.
- In the lower right of the Custom Fields area, click **Define Fields**.
- In the Define Fields window, under the Label column, enter the field names you want to create. See Figure 14-11.
- In the columns to the right, check the boxes to select who you're using the custom field for.
- Click **OK** in the Define Fields window.
- Click **OK** in the Edit Vendor window to create the field.

Figure 14-10
Use the Additional Info tab on the Edit Vendor window to create custom fields.

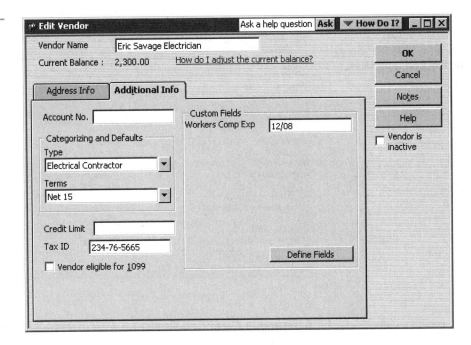

Figure 14-11
Use the Define Fields window to name a custom field and select who the field will be used for.

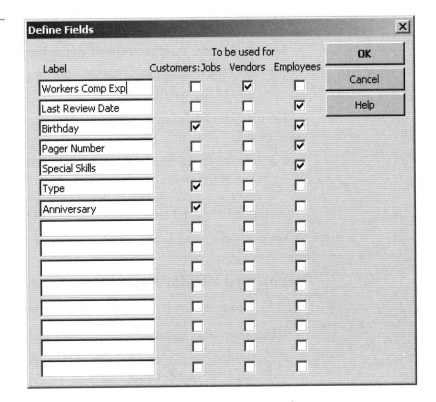

Figure 14-12
The Templates list shows the existing templates.

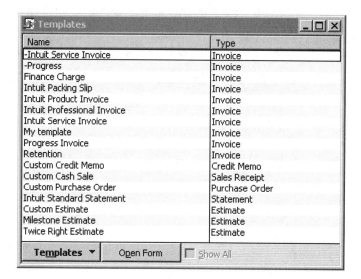

Adding a Custom Field to Purchase Order Forms

To add a custom field to your purchase order forms:

- From the **Lists** menu, choose **Templates**. In Figure 14-12, we show the Templates list.

- Click **Custom Purchase Order** in the **Name** column.

- Choose **Edit Template** from the pull-down **Templates** menu at the bottom of the window.

- Click the **Fields** tab to get a window like the one shown in Figure 14-13.

- Select the check box next to **Workers Comp**. You can also specify if you want the custom field printed on the purchase order, if you want it to appear only on the screen, or if you want both. Usually you'll choose to see the custom field only on the screen.

- Click **OK** to add the field to the purchase order form.

Preparing a Vendor Report on a Custom Field

Here's how to make a vendor report on a custom field:

- From the **Lists** menu, choose **Vendor List**.

- On the Vendor List window, click **Reports** at the bottom of the window and choose **Contact List**.

- On the **Display** tab of the Modify Report: Vendor Contact List window, under the **Columns** box, scroll through the list to select only **left margin**, **Vendor**, **Phone**, **Contact**, and **Workers Comp Exp**. See Figure 14-14.

Chapter 14: Payables

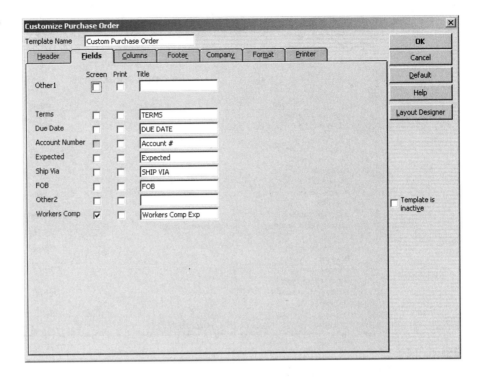

Figure 14-13
Use the Fields tab window to select which field you want to see in the Purchase Order window and which fields you want printed on your purchase orders.

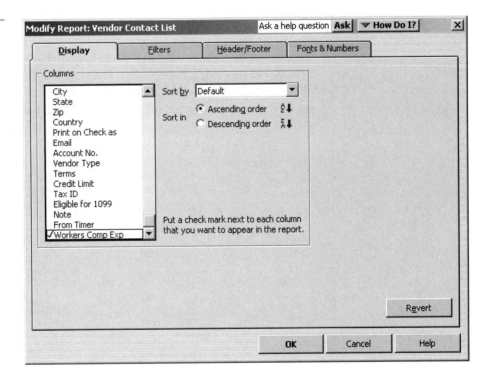

Figure 14-14
Use the Display tab of the Modify Report: Vendor Contact List window to select the columns you want to appear in a Vendor Contact list.

Chapter 14: Payables

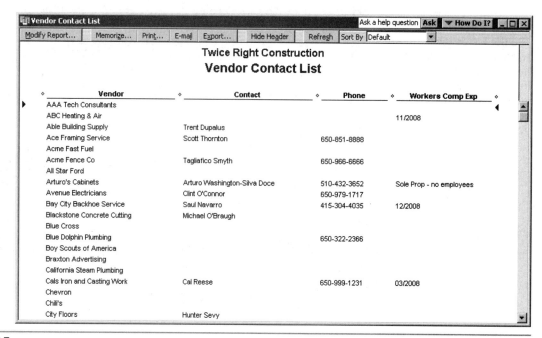

Figure 14-15
This Vendor Contact List report shows the custom field Workers' Comp Expires.

> Click **OK** to see a report like the Vendor Contact List we show in Figure 14-15.
>
> Now that you've finished entering bills and writing checks, you can proceed to payroll, Chapter 15. Even if you don't have any employees, you should read through this chapter because we've included a special section on how to enter time for sole proprietors and partnerships.

Chapter 15

Payroll

QuickBooks Pro's ability to handle payroll is one of its best features. It has tremendous flexibility, which is also why setting it up can be somewhat frustrating.

In this chapter we'll show you how to:

- enter a timesheet
- process payroll
- print payroll checks

And, if your company is a sole proprietorship or partnership, we'll show you how to enter and process owner's time to jobs so it shows on your job cost reports. But before we begin, make sure you have all your payroll items set up. You'll find information on payroll items back in Chapter 5.

Entering a Timesheet

Before creating a payroll check for any employee, you need to complete a pay period timesheet to assign the hours the employee worked to a customer, job, service item (job phase), and payroll item. You get this information from the weekly timesheet filled out by the employee, foreman, or supervisor in charge of the project. Some states require that employees complete their own timecards.

Before you process payroll the first time, we suggest you open the *sample.qbw* file that comes with this book. We'll walk you through entering a weekly timesheet and processing payroll. The employee we'll be using first is Joe Bliss. In our sample company world, he's filled out and turned in a timecard for the week ending December 27, 2009. To begin, we're going to enter the information on his timecard into QuickBooks Pro.

Chapter 15: Payroll

To enter a timesheet:

- From the **Employees** menu, choose **Time Tracking**, and then **Use Weekly Timesheet**.

- From the **Name** drop-down list, select the employee name. Figure 15-1 shows our Weekly Timesheet example for employee Joe Bliss.

You'll see the current week in the upper left-hand corner of the window. You have to make sure you're entering the time for the correct week. To move to different weeks:

- Click **Previous** or **Next**. To search for a particular date, click **Set Date** and enter the date you want in **New Date**. In our example, it's December 21, 2009.

Let's look at the Weekly Timesheet in Figure 15-1 in detail.

The Customer:Job column represents the customer and job. As you see, Joe Bliss worked on the Rachel Cohn:Guest Cottage.

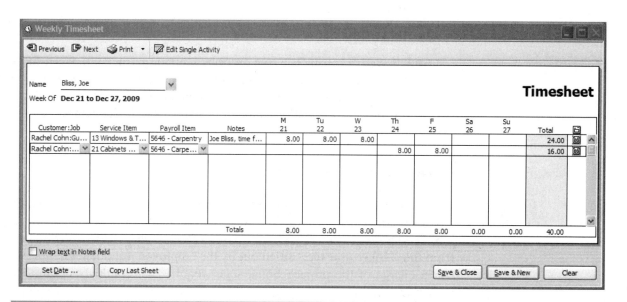

Figure 15-1
The QuickBooks Pro timesheet assigns the hours worked to a Customer:Job, Service Item, and Payroll Item.

Figure 15-2
Warning window for a payroll item that hasn't been set up.

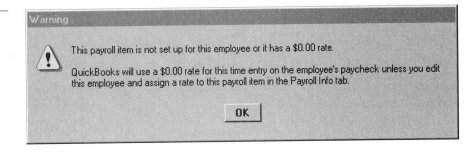

The Service Item column represents the job phases the employee worked on. For more information on job phases, see Chapter 4, Items. In our example, Joe worked on different job phases during the week. On Monday through Wednesday he worked on Rachel Cohn:Guest Cottage installing windows and trim. Then on Thursday and Friday Joe installed cabinets.

The contents of the Payroll Item column depend on whether or not you're using payroll items to track workers' compensation. In our example, we use workers' compensation codes as Payroll Items in QuickBooks Pro. This lets us print a report later to help us fill out our workers' compensation report.

Entering a Payroll Item Not on an Employee's Payroll Info Tab

If you enter a payroll item that isn't on the employee's Payroll Info tab, you'll get a warning like the one shown in Figure 15-2. The Payroll Info tab is a part of an employee's record. To fix this problem:

- From the **Lists** menu, choose **Employee List**.
- In the Employee List window, click the employee you're entering the timecard for.
- From the pull-down **Employee** menu, choose **Edit** and then, in the **Change Tabs** Field, choose **Payroll and Compensation Info**.
- In the Earnings section of the Edit Employee window, click in the **Item Name** column and select the item(s) you need from the pull-down Payroll Item list. Our example in Figure 15-3 shows the Payroll Info tab for Joe Bliss.
- Click **OK**.

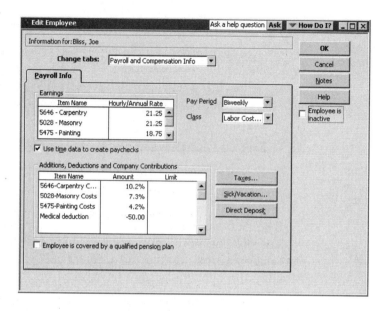

Figure 15-3
Payroll items for an employee are listed in the Earnings box of the employee's data record.

Returning to the Weekly Timesheet, the Notes column is extremely important if you'll be billing a customer for an employee's labor on a time-and-materials basis. If you use QuickBooks Pro to generate an invoice for the customer, make sure you fill in Notes with the exact text you want the customer to see on the invoice. The text you enter here will appear as a description on the invoice.

In the box to the right of the Total column, you'll see an icon that represents an invoice. You click this icon if you don't want the hours on the timesheet to be kept for time-and-materials invoicing. If you click the icon, you'll see a red X on the icon to tell QuickBooks Pro the time's not billable on a time-and-materials basis for this job.

Now back to what to enter on the timesheet:

- Enter a different line for the hours worked on each job, phase, and (if needed) workers' comp code.
- Enter the time for the week.
- Click the invoice icons as necessary for the jobs.
- Click **Save & Close** to record the timesheet.

Don't forget that any report you generate before you process payroll will show only the hours you've entered, not the dollar amounts for those hours. After you process payroll, the dollar amounts associated with the hours will appear on your reports.

Processing Employee Payroll

Before you can process any payroll for an employee, you must enter a timesheet for the employee. And before you can process payroll for the first time you need to go through the steps outlined below to set up the payroll module. To process payroll the first time for employees you've entered timesheets for:

- From the **Employees** menu, choose **Pay Employees**.
- If you're working with a new QuickBooks Pro file, you'll be prompted to select a payroll service. Click **Yes**.
- Select **Choose a payroll option**. See Figure 15-4.

At the Choose a payroll option window, you have the following options (see Figure 15-5):

1. *QuickBooks Standard or Enhanced Payroll* – Replaces the Tax Table Service. This option provides you with up-to-date federal and state tax tables and federal forms. For more information about this service, contact Intuit at 800-365-9618.

2. *Assisted Payroll* — Offers the Standard or Enhanced Payroll features, plus payroll tax depositing and filing, W-2 printing and mailing, and a "no penalties" guarantee.

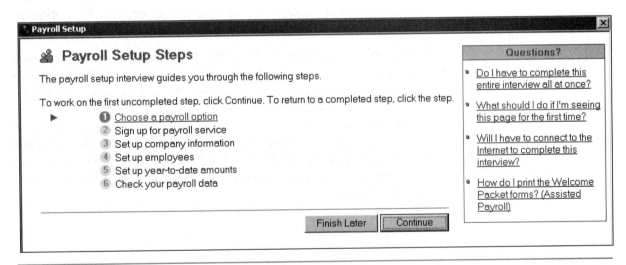

Figure 15-4
In the Payroll Setup window, select Choose a payroll option.

Chapter 15: Payroll

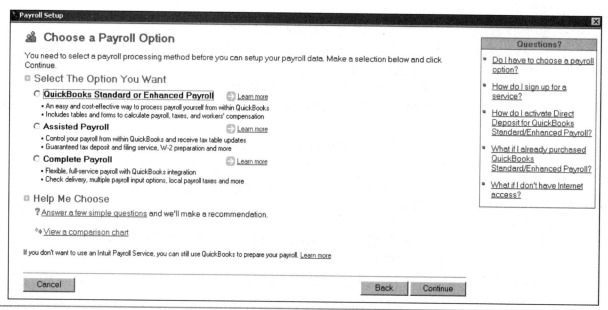

Figure 15-5
If you have a payroll service and want to get your labor cost to jobs, you need to use the QuickBooks Pro payroll module. If you select *Learn more*, you don't have to pay for the payroll tax tables.

3. *Complete Payroll* — An Intuit outside payroll service that integrates with QuickBooks Pro. This option offers employee check printing, tax deposits, tax reporting, direct deposit, W-2 printing, and a "no penalties" guarantee.

4. *Manual Calculations* — To locate this option, select *Learn more* after the sentence: *If you don't want to use an Intuit payroll service, you can still use QuickBooks to prepare your payroll.* See Figure 15-5. At the Payroll Setup window, select the button: *I choose to manually calculate payroll tax.* If you select this option, you need to manually calculate the payroll taxes for each employee each time you run a payroll check. Select this option if you have a payroll service and just want to use QuickBooks Pro payroll to get labor costs to job cost reports.

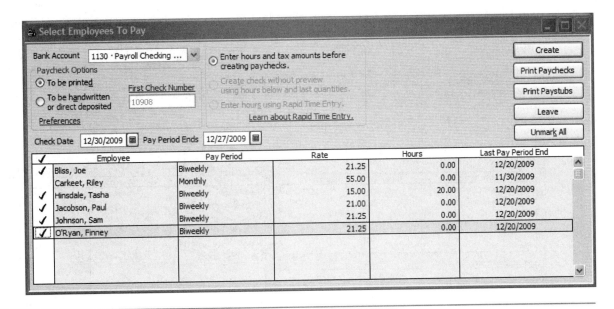

Figure 15-6
Put a check mark next to each check you're ready to print.

From the **Employees** menu, select **Pay Employees**.

- In the Select Employees To Pay window, select the correct Bank Account you'll use for the checks. In our example shown in Figure 15-6, we used our 1130 Payroll Checking Account.

- Enter **Check Date** and **Pay Period Ends** date.

If hours don't show up for an employee even though you know you've entered time for that employee, you probably didn't enter the pay period ending date correctly. To make sure all time you entered on a previous timesheet shows up, you must enter a date after the last date that shows up on the previous timesheet you entered. For example, in Figure 15-1, the last date on the timesheet next to Week Of is 12/27/2009. To pick up all the time on that timesheet, the date you enter into the Pay Period Ends box must be 12/27/2009. Also, the date must not conflict with the next timesheet you enter.

- Click in the leftmost column next to the employee name for each employee you're paying.

- Click **Create**. A Payroll Subscription warning box may appear, depending on your system and tax tables.

Chapter 15: Payroll

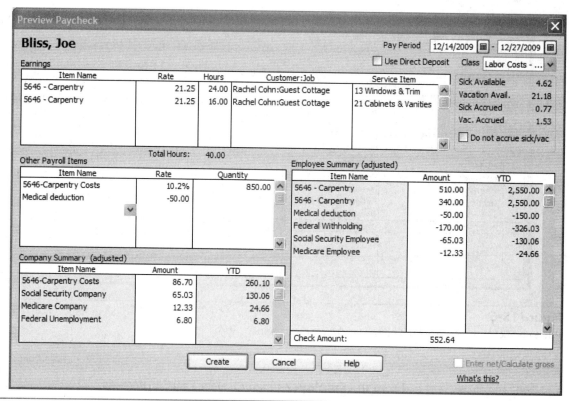

Figure 15-7
Use the Preview Paycheck window to make sure you enter all costs for an employee correctly.

Using the Preview Paycheck Window

Now you'll see the Preview Paycheck window that will display the information you entered from the timecard. The hours for each payroll Item, Customer:Job, and Service Item you entered will show in the Earnings box. The information in the Earnings box comes directly from the timecard information you input onto the Weekly Timesheet. Our example shown in Figure 15-7 is the Preview Paycheck window we get using the timesheet shown in Figure 15-1.

You must make sure here that the information in the Earnings box is correct. If any information isn't correct, you should click Cancel to close the Preview Paycheck window right away. Then return to the employee's timesheet you entered and make the necessary changes. Also, if you make changes in the Earnings box in the Preview Paycheck window, the timecard won't be automatically changed. You must make any corrections directly on the timesheet.

198 Contractor's Guide to QuickBooks Pro

Entering Payroll Additions and Deductions

You use the Other Payroll Items box to track additions to an employee's pay (bonus), deductions from an employee's pay (medical or dental), and any expenses that the company has because of the employee's paycheck (workers' compensation expense). In our example, we deducted $50 from Joe's paycheck for medical insurance. To do this, in the Preview Paycheck window:

- Click in the **Item Name** column under the **Other Payroll Items,** box and select the addition or deduction from the pull-down Item Name List.

- In the **Rate** column, enter the amount of the addition or deduction. Check to make sure you've entered the information correctly. To enter a deduction, enter a negative number in the **Rate** column. In our example in Figure 15-7, we deducted $50 for medical insurance from Joe Bliss's check.

- Be sure to check your entries in the Employee Summary box.

If you want to send the cost of workers' compensation to the Customer:Jobs listed in the Earnings Box, you'll need to enter the workers' compensation codes (company contributions payroll items) under the Other Payroll Items box. If you haven't set up these payroll items, you'll find information on how to do this back in Chapter 5.

You also use the Preview Paycheck window to add the cost of workers' compensation to jobs. To do this:

- In the **Item Name** column, under the **Other Payroll Items** box, select the workers' compensation code that matches the first line listed in the Employee Summary box. In our example, Joe has 5646-Carpentry Costs listed as the first line in the Employee Summary box. We select its matching company contribution payroll item in the Item Name column, 5646-Carpentry Costs. The rate will appear automatically, but you'll need to fill in **Quantity**. In our example, Quantity is the total of all wages 5646-Carpentry Costs in the **Amount** column of the Employee Summary box. 5646-Carpentry Costs appears two times. So Quantity is $510 + $340 which is $850.

- Continue entering a matching workers' compensation code for each payroll item listed in the Employee Summary box that's subject to workers' comp.

Chapter 15: Payroll

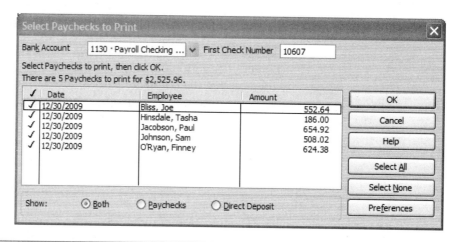

Figure 15-8
Put a check mark next to each check you're ready to print.

If you follow this procedure, the cost of the workers' compensation insurance will be spread to each of the jobs listed in the Earnings box.

Finally, check the employee's information on the Preview Paycheck window again. If everything is correct:

▌ Click **Create**.

When you've completed creating the check:

▌ Click **Leave** in the Select Employees To Pay window.

The payroll check you created will be entered in the checking account you selected.

Printing Employee Checks

To print the employee checks:

▌ From the **File** menu, choose **Print Forms** and then **Paychecks**.

▌ Select the paychecks to print, as shown in our example in Figure 15-8.

▌ Click **OK** to print the checks.

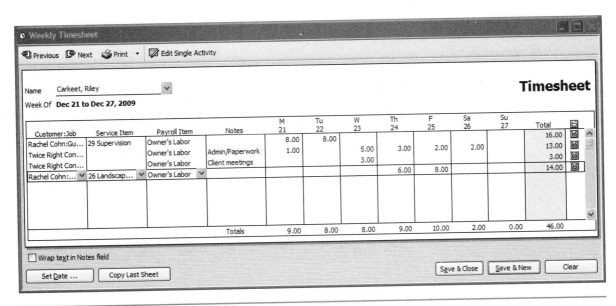

Figure 15-9
When entering a timesheet for an owner, the payroll item must be Owner's Labor.

Allocating Sole Proprietor or Partner's Time to a Job

One of the most common questions we hear is "How do I, as an owner, allocate my time to a job?" All you have to do is enter your time on a timecard and then process payroll. While you process payroll, you deduct the dollars back out by creating a zero check for yourself. Entering the timecard and processing payroll will add the hours, and the dollars associated with those hours, to your job cost reports. But be sure you've set up payroll items for owner's wages and owner's labor deduction. See Chapter 5 if you need help setting up a payroll item.

Entering a Timesheet for an Owner

For each pay period, enter the owner's time on a timesheet just as you would a regular employee. Let's go through an example for Riley Carkeet, the sole proprietor of Twice Right Construction. To enter a timesheet for an owner or partner:

- From the **Employees** menu, choose **Time Tracking**, and then **Use Weekly Timesheet**.

- From the **Name** drop-down list, select the owner. Figure 15-9 shows our Weekly Timesheet example for Riley Carkeet.

- Select the appropriate week. In our example, it's December 21 - December 27, 2009.
- Enter a different line for the hours worked on each job and phase.
- In the **Payroll Item** column of each line on the timesheet, make sure you select the payroll item **Owner's Labor**. You must always use this item when you enter owner's time. The Owner's Labor item is set up to post to an equity account (account 3999 Owner's Time to Jobs in our *sample.qbw* file). If you select a payroll item that doesn't post to the appropriate equity account, the owner's labor costs won't appear properly in your job cost reports.
- Click the invoice icons as necessary for the jobs.
- Click **Save & Close** to record the owner's time.

Processing an Owner's Time

To process payroll allocating owner's time that you have entered a timesheet for:

- From the **Employees** menu, choose **Pay Employees**.
- In the Select Employees To Pay window, select **Enter hours and preview check before creating.**
- From the pull-down Bank Account list, select your **Adjustment Register** account. Since you're creating a zero check, you don't want it showing up in the regular bank account. Use the Adjustment Register to hold transactions with zero amounts. In our example shown in Figure 15-10, we use 1111 Adjustment Register account.
- Enter the **Check Date** and **Pay Period Ends**.
- Click in the checkmark column next to the owner's name.
- Click **Create**.

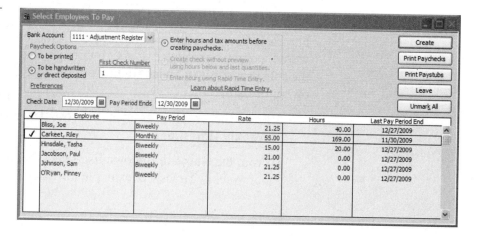

Figure 15-10
Use your Adjustment Register account when you process an owner's paycheck.

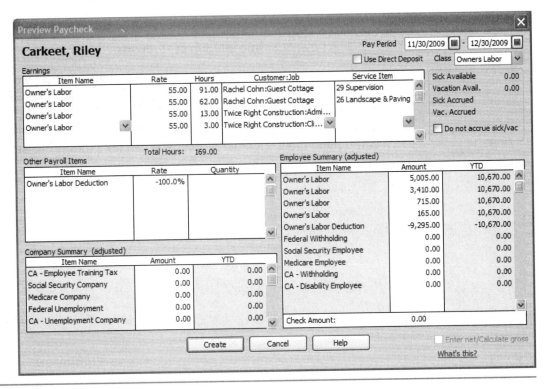

Figure 15-11
When processing payroll for an owner, make sure the Owner's Labor Deduction and rate are entered correctly. It's important that the Check Amount is 0.00.

Now you'll see the Preview Paycheck window for the owner's timesheet. Figure 15-11 shows our example for Riley Carkeet. You want to be sure you get Owner's Labor Deduction in the Other Payroll Items box in this window. To do this:

- In the **Item Name** column, under the **Other Payroll Items** box, select the payroll item **Owner's Labor Deduction**.

- In the **Rate** column enter –100%. See Figure 15-11.

- Make sure **Check Amount** is 0.0 when you finish entering these amounts.

- Click **Create** to record the check.

In this chapter, we gave you a great deal of information. We told you how to:

- enter timesheets
- process payroll

- allocate workers' comp costs to jobs
- get owner's time into job cost reports

If you feel a bit overwhelmed by all the information we've presented, keep in mind that you don't have to take all of this in at one time. For example, you can simply process payroll first and then add workers' comp costs later. Then you can add owner's time when you have payroll and workers' comp figured out. In other words, you can take things one step at a time.

Chapter 16

Using QuickBooks Pro on a Cash Basis

You can use QuickBooks Pro on a cash basis or an accrual basis. Or, you can start on a cash basis and later change to an accrual basis. All you have to do is change one option in your preferences.

In the previous two chapters we showed you how to do your receivables and payables on the accrual basis. That's what most contractors will use. But if you're a small operator with few or no employees, or no time to do bookkeeping, you're probably running your business on a cash basis.

If you're just starting your business or have just made the transition from Quicken, you may feel more comfortable starting to use QuickBooks Pro on a cash basis. Although this may be an easier way to get up and running, we suggest that you try to change to the accrual method as soon as possible. Accrual-based reports will give you a more timely and accurate picture of your finances, job status, and operating status. Since QuickBooks Pro has the ability to switch reporting between cash and accrual, your accountant can still file your tax returns on a cash basis, and you can get your management reports on the more accurate accrual basis.

You may decide to use the cash basis of accounting if:

- you're just starting your business and know "off the top of your head" what the payables and receivables are.

- you've just started to get computerized and are overwhelmed by getting everything into the computer.

- you have very little time to do bookkeeping.

So let's look at the cash basis of accounting. To report an expense, you simply write a check and log it in. To report income, you deposit money in the bank and log in the deposit. It's easy and fast. You don't have to enter a bill or invoice and then tell the computer it has been paid. The downside, of course, is that the computer isn't tracking who you owe money to, or who owes you money. Also, your reports don't include expenses that you haven't paid or invoices you haven't issued to customers.

Chapter 16: Using QuickBooks Pro on a Cash Basis

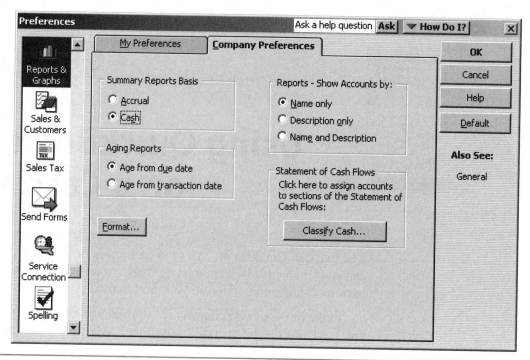

Figure 16-1
Use the Reports & Graphs Preferences window to let QuickBooks Pro know you operate on a cash basis.

To tell QuickBooks Pro that you want to operate on a cash basis:

- From the **Edit** menu, choose **Preferences**.
- Click the **Reports & Graphs** icon. See Figure 16-1.
- Click the **Company Preferences** tab.
- Select **Cash** for the Summary Reports Basis.
- Click **OK**.

How to Record a Check

To record a check:

- From the **Banking** menu, choose **Write Checks**.

Now let's look at Figure 16-2 to see what you would need to enter here:

- In **Bank Account**, make sure you select the correct bank account.
- Check **To be printed** unless you're recording a handwritten check.
- In **No.**, enter the check number if it's a handwritten check.

206   Contractor's Guide to QuickBooks Pro

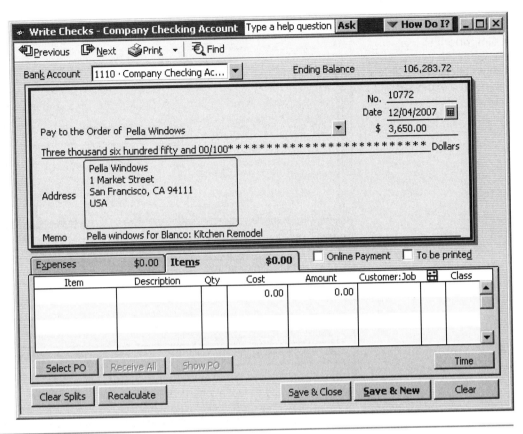

Figure 16-2
The QuickBooks Pro Write Checks window looks similar to a regular paper check.

- In **Date**, enter the date of the check.
- In **Pay to the Order of**, enter payee name or select the name from the drop-down vendor list.
- In **$**, enter the check amount. The Dollars field will fill in automatically.

Using the Items Tab to Record a Job-Related Cost

You should always use the Items tab to enter job-related costs. Our example in Figure 16-3 shows the Items tab on a check to Pella Windows for the job-related cost of windows for Jennifer Blanco:Kitchen Remodel.

To record a check for a job-related cost in a Write Checks window:

- Click the **Items** tab. Click in the **Item** column and select the appropriate items from the drop-down list.
- In **Qty**, enter the quantity of the item. Amount will be calculated automatically.

Chapter 16: Using QuickBooks Pro on a Cash Basis

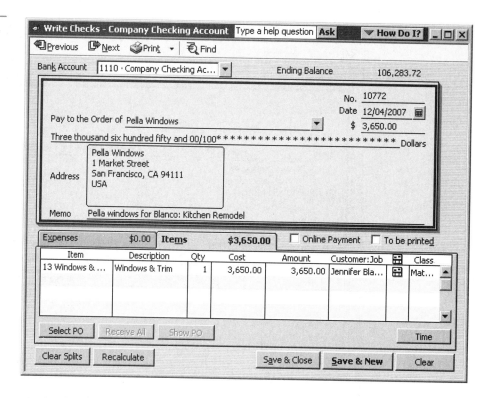

Figure 16-3
It's important to use the Items tab when you write checks for job-related expenses.

- Select **Customer:Job** from the pull-down Customer:Job list.

- Click the invoice icon so there's a red **X** over it if you're *not* billing the check to the client. If you are, don't click the invoice.

- In **Class**, select the appropriate class from the pull-down Class list (if you're using classes).

If you're splitting the total payment to the vendor over more than one job (or if the payment covers more items), click in the next line and make the additional entry(s) using the same procedure.

- Click **Save & Close**.

Using the Expenses Tab to Record an Overhead Expense

You should always use the Expenses tab to enter overhead costs. Our example in Figure 16-4 shows check number 10774 to Acme Fast Fuel for gasoline, which is an overhead expense.

To enter an overhead expense check in a Write Checks window:

- Click the **Expenses** tab.

- Click in the Account column and select the correct expense account from the drop-down list. In our example, we used 6101 - Gas & Oil.

Figure 16-4
Use the Expenses tab when you write checks for overhead expenses.

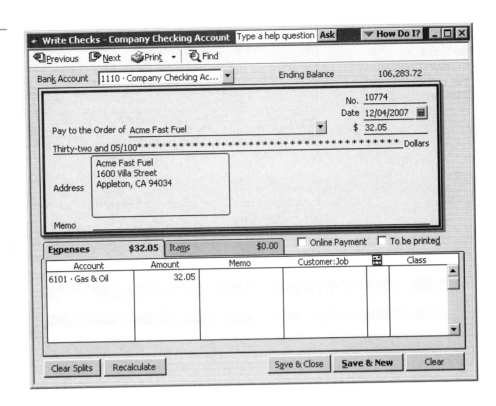

If you're splitting the total payment to the vendor over more than one expense account, click in the next line and make the additional entry(s) using the same procedure.

When the check entry is complete:

▌ Click **Save & Close** to record the check.

How to Record a Deposit

You record a deposit to your checking account in the Make Deposits window. To record a deposit:

▌ From the **Banking** menu, choose **Make Deposits**.

▌ In the Payments to Deposit window, select the payments you want to deposit.

▌ Click **OK**.

Look at Figure 16-5 to see what you would need to enter here:

▌ In the Make Deposits window, in **Deposit To**, select your checking account from the pull-down list. If you have multiple checking accounts, make sure you select the correct one.

Chapter 16: Using QuickBooks Pro on a Cash Basis

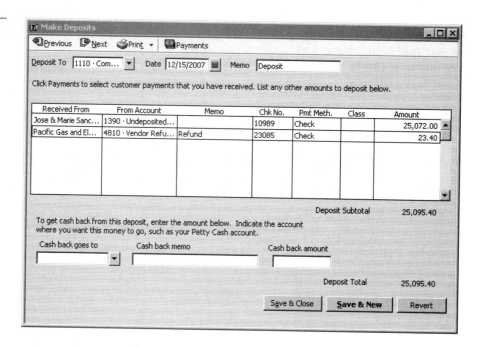

Figure 16-5
The Make Deposits window shows all checks that are ready to be deposited.

- In **Date**, enter the date of the deposit.
- In **Received From**, use the pull-down list to select who you've received the check from.
- In **From Account**, select the correct account for the deposit from the pull-down list. Usually this is an income account.
- In **Memo**, enter a short note about the deposit you're recording.
- In **Chk No.**, enter the check number.
- In **Pmt Meth.**, select payment method from the pull-down list.
- In **Class**, select the correct class for the deposit from the pull-down list. Usually, you'll use Revenue - Job Related. If the deposit isn't job related, you can skip this.
- In **Amount**, enter the deposit amount.
- Click **Save & Close** to record the deposit.

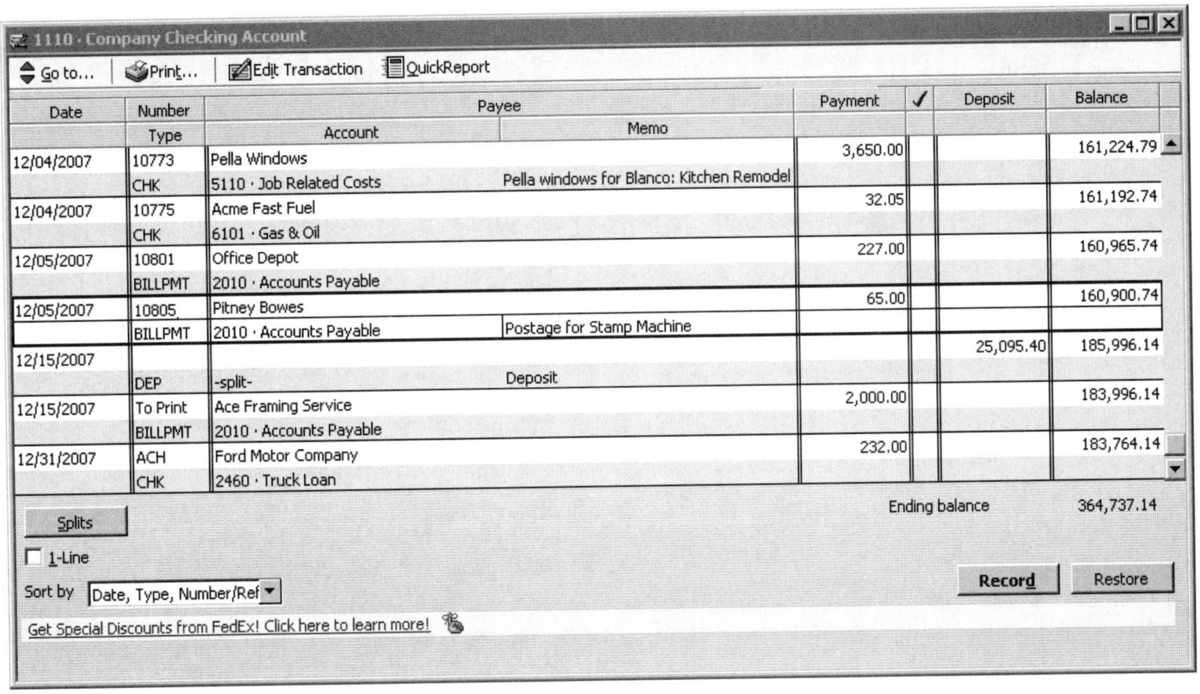

![Figure 16-6 register screenshot]

Figure 16-6
Your checking account register shows all your payments, deposits, and your account balance.

Checking Your Transactions with the QuickBooks Pro Register

QuickBooks Pro has an on-screen register you can use to look at all your recorded transactions. It displays all of your recorded transactions in a list. For example, you can use the register to get a quick look at all the checks you've written over a certain time. Or you can use it to get your checkbook account balance.

To open the Checking Register:

- From the **Lists** menu, choose **Chart of Accounts**.

- In the Chart of Accounts list, double-click your checking account to bring up the register. See Figure 16-6.

This register will show you payments, deposits, and current balance of your checking account.

Now that you've learned how to enter transactions on a cash basis, keep in mind that running your construction company on a cash basis isn't the best way to run the business. As soon as possible, you should switch to an accrual basis of accounting so your job cost reports will always be up to date.

Chapter 17

Reports

As a contractor, you need up-to-the-minute information on job costs — but "up-to-the-minute" means different things to different contractors. If you typically run jobs that last only a week or so, you'll probably want to analyze your job costs daily. That's the only way to keep up with them. If your jobs run months or years, you don't need to get job cost reports so frequently. Once a month may be sufficient for longer jobs.

In this chapter, we'll look at reports for the construction business on job costing. We'll cover:

- how to customize existing reports.
- how to use the memorized reports in the *sample.qbw* and *company.qbw* data files on the CD-ROM that comes with this book.
- how to use the project reports in QuickBooks Pro.

Finding and learning to use reports will give you more insight into the day-to-day status of your jobs. Learning how to modify reports will give you the power to create reports that are meaningful to you and more specific to your business.

We added the memorized reports in *sample.qbw* and *company.qbw* to give you reports you can use on a day-to-day basis. We've customized the reports for the construction industry.

How to Modify Reports

You can use the Modify Report window to pick the information you want to include in the report. For example, you may want to get a report for just one job, or you may want to change a report to include an invoice or check number. When you display a report on your computer, you'll see the Modify Report window blocking the report. To view the report, click the cancel button on the Modify Report window. To retrieve the Modify Report window, click the button titled Modify Report in the top left corner of the report window. You can see this button in Figure 17-1, where we show a Job Profitability Detail report.

Chapter 17: Reports

```
Twice Right Construction
Job Profitability Detail for Jennifer Blanco:Kitche...
                    All Transactions
                          Act. Cost    Act. Revenue    ($) Diff.
Service
  01 Plans & Permits
    01.1 Plans              172.40         0.00        -172.40
    01.2 Building Permits   243.00         0.00        -243.00
  Total 01 Plans & Permits  415.40         0.00        -415.40

  06 Floor Framing        6,000.00         0.00       -6,000.00
  07 Wall Framing        60,000.00         0.00      -60,000.00
  08 Roof Framing        10,000.00         0.00      -10,000.00
  11 Siding               4,000.00         0.00       -4,000.00
  12 Doors & Trim         5,000.00         0.00       -5,000.00
  13 Windows & Trim       7,862.02         0.00       -7,862.02
  14 Plumbing            12,500.00         0.00      -12,500.00
  16 Electrical & Lighting 5,500.00        0.00       -5,500.00
  18 Interior Walls       1,500.00         0.00       -1,500.00
  21 Cabinets & Vanities 24,374.68         0.00      -24,374.68
  22 Specialty            9,000.00         0.00       -9,000.00
  23 Floor Coverings     15,000.00         0.00      -15,000.00
  24 Paint                  300.00         0.00         -300.00
  29 Supervision              0.00         0.00            0.00
  Fixed Price Billing         0.00   224,000.00     224,000.00
Total Service           161,452.10   224,000.00      62,547.90

TOTAL                   161,452.10   224,000.00      62,547.90
```

Figure 17-1
To change a report, click the Modify Report button located in the top left corner of the report window.

The Display Tab on the Modify Report Window

In general, you use the Display tab of a Modify Report window to select the date range and the columns you would like to see on a report. Figure 17-2 shows the Display tab column settings for the Job Profitability Detail report in Figure 17-1. It's worth noting here that the Display tab options change depending on the report you're modifying. For example, in Figure 17-2 you can only select from the four columns shown in the Display tab window — Act. Cost, Act. Revenue, $ Difference, and % Difference. In Figure 17-3, we show the Display tab of the Modify Report window for a Customer Phone List report. There are many more options for this report.

214 Contractor's Guide to QuickBooks Pro

Figure 17-2
The Display tab window lets you specify which dates and columns will be on the report.

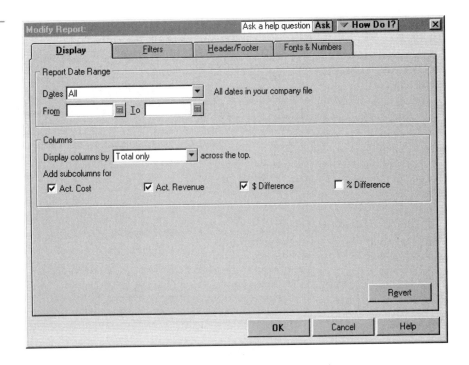

Figure 17-3
The Display tab window changes depending on the type of report you're generating.

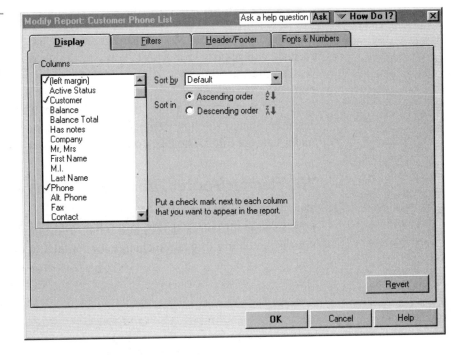

Figure 17-4
The Filters tab window lets you pick which data is on your report.

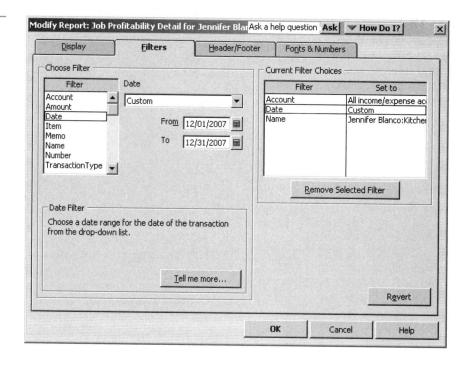

The Filters Tab on the Modify Report Window

The Filters tab on the Modify Report window lets you specify which specific records will appear in a report. You can define the accounts, customers:jobs, classes, transaction types, payees, dates, transaction amounts, and/or memo contents that each record must match before it's included in the report. In Figure 17-4, we show a Filters tab window to get a report for a particular job — Jennifer Blanco:Kitchen Remodel. At the right side of the Filters tab window under Current Filter Choices, you'll see a list of the current filters for the report.

The Header/Footer Tab on the Modify Report Window

The Header/Footer tab on the Modify Report window lets you customize your report and modify your header and footer information. When memorizing new reports, you'll use this tab to change the report's title.

Figure 17-5
The Memorized Report List in *sample.qbw* and *company.qbw*.

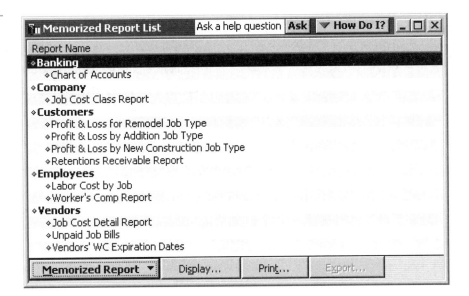

Using Our Memorized Reports

Memorizing a report is easy — but creating the report the first time may take some time. Determining exactly what combination of modifying and filtering will give you the report you want can be a bit of a brain twister. Also, you have to be absolutely consistent and accurate when you enter transactions. If you enter a record incorrectly, it may not be in the reports when, or where, it should be. You don't want to make a good decision based on bad information.

We defined and memorized the reports we discuss in this section in the *company.qbw* and *sample.qbw* data files included on the CD that comes with this book. We hope you can make good use of these reports in managing your business. We based the reports on the *sample.qbw* and *company.qbw* data files. If you've changed these files or are using other data files, you'll need to review the reports to make sure they show what you need.

However, you may have to change the dates used in the memorized reports. You can use the methods discussed in the previous sections to do that.

To display a list of the memorized reports in *sample.qbw* or *company.qbw*:

▌ Open **sample.qbw** or **company.qbw**.

▌ From the **Reports** menu, choose **Memorized Reports** and then **Memorized Report List**. See Figure 17-5.

▌ To generate a memorized report, select the report and click **Display**.

Figure 17-6
A memorized Chart of Accounts Report.

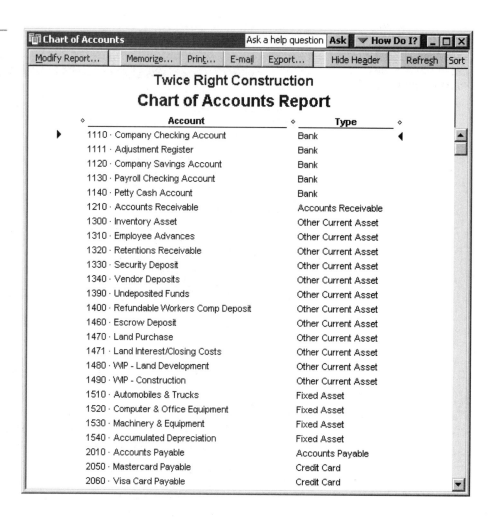

If you didn't use the *sample.qbw* or *company.qbw* data file to set up your QuickBooks Pro company file, you may want to memorize some of the reports in this section in your own company file. We'll tell you how to do that after we describe each report.

However, if you did use our *sample.qbw* or *company.qbw* data file to set up your company data file, you can just skip the information on memorizing each report.

Now let's look at each memorized report.

Chart of Accounts Report

The memorized Chart of Accounts Report simplifies the QuickBooks Pro built-in Chart of Accounts Report. It looks similar to the one we printed and included in Chapter 3. Use this report to help define which accounts you have set up in your Chart of Accounts. Figure 17-6 shows a memorized Chart of Accounts Report.

Figure 17-7
Choose the columns to appear in the Chart of Accounts Report.

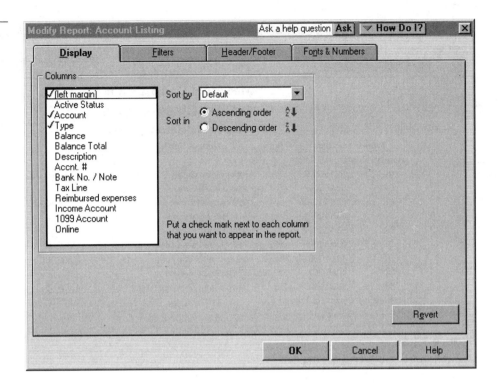

Memorizing the Chart of Accounts Report — To memorize the Chart of Accounts Report in your QuickBooks Pro data file:

- From the **Reports** menu, choose **List**, then **Account Listing**.
- Now select the columns in the Display tab of the Modify Report window, as shown in Figure 17-7. Select only **(left margin)**, **Account**, and **Type** from the Columns list. Be sure to scroll down the list to deselect other columns on the list.
- Click **OK**.
- In the report window, click **Memorize**.
- In the **Memorize Report** window, enter **Chart of Accounts**.
- Click **OK**.

Chapter 17: Reports

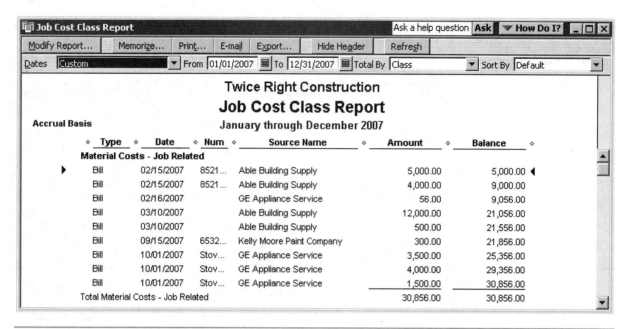

Figure 17-8
A memorized Job Cost Class Report.

Job Cost Class Report

The memorized Job Cost Class Report shows you, in detail, all revenue and costs separated by class. This report is a Transaction Detail by Account report sorted by these classes — Revenue, Labor, Materials, Subcontractor, Equipment Rental, and Other. The bold column header at the top left of each section is the class. In the example we show in Figure 17-8, it's Material Costs - Job Related.

Memorizing the Job Cost Class Report — To memorize the Job Cost Class Report in your QuickBooks Pro data file:

- From the **Reports** menu, choose **Accountant & Taxes**, and then **Transaction Detail by Account**.
- Select the columns in the Display tab of the Modify Report window, as shown in Figure 17-9. Select only **(left margin), Type**, **Date**, **Num**, **Source Name**, **Amount**, and **Balance**. Scroll down to the bottom of the list to find Amount and Balance.
- In **Report Date Range**, enter the date range you're interested in. Usually this would be for a given month or year. Our example in Figure 17-9 uses the fiscal year 1/1/2007 to 12/31/2007.
- In **Total By**, select **Class** from the pull-down list.
- Click **OK**.
- In the report window, click **Memorize**.

220   *Contractor's Guide to QuickBooks Pro*

Chapter 17: Reports

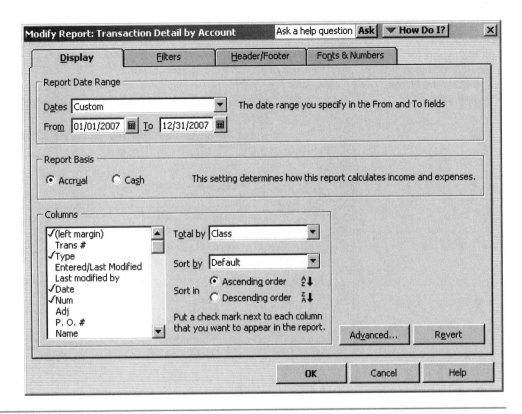

Figure 17-9
In the Total by field, select Class to group and subtotal transactions by job cost class.

■ In the **Memorize Report** window, enter **Job Cost Class Report**.

■ Click **OK**.

Often you'll want to know the job-related labor costs for just one job. To do this:

■ Click the **Modify Report** button when you have a Job Cost Class report on your screen.

■ Click the **Filters** tab.

■ In the Filters tab window, select **Name** from the Filter list. See Figure 17-10.

■ From the Name drop-down list, select the **customer:job** you're interested in.

■ Click **OK**.

Contractor's Guide to QuickBooks Pro 221

Chapter 17: Reports

Figure 17-10
To filter a report for a specific job, select the job from the Name drop-down list.

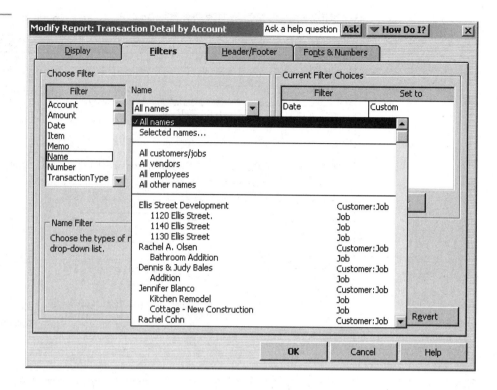

Retentions Receivable Report

The Retentions Receivable Report lists all the customers that owe you for retention for the dates you specify. See Figure 17-11.

Memorizing the Retentions Receivable Report — To memorize the Retentions Receivable Report in your QuickBooks Pro data file:

- From the **Reports** menu, choose **Customers & Receivables**, and then **Customer Balance Summary**.

- In **Report Date Range**, enter the date range you're interested in. Usually this would be for a given month or year.

- Click the **Filters** tab.

- In the Filters tab window, select **Account** from the Filter list.

- From the pull-down Account list, select your Retentions Receivable account. In our example shown in Figure 17-12, the account is 1320 - Retentions Receivable.

- Click **OK**.

- In the report window, click **Memorize**.

- In the **Memorize Report** window, enter **Retentions Receivable Report**.

- Click **OK**.

222 Contractor's Guide to QuickBooks Pro

Chapter 17: Reports

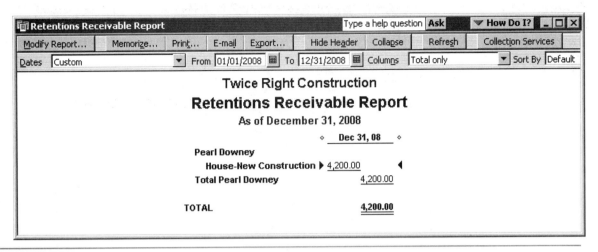

Figure 17-11
A memorized Retentions Receivable Report.

Figure 17-12
The Report Filters window showing some of the settings for the Retentions Receivable Report.

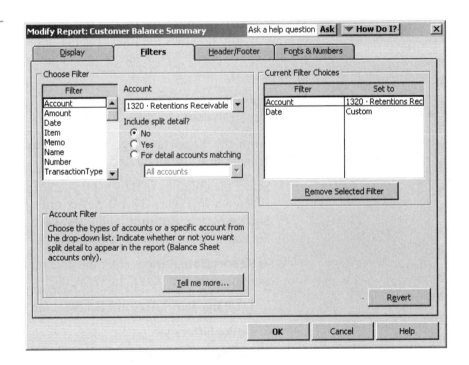

Contractor's Guide to QuickBooks Pro **223**

Figure 17-13
A memorized Labor Cost by Job report.

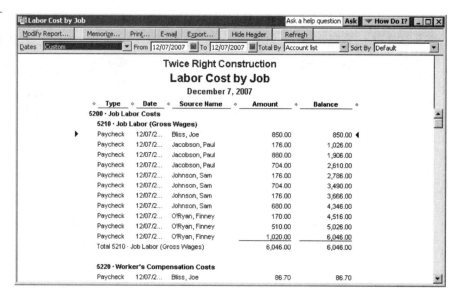

Labor Cost by Job Report

The memorized Labor Cost by Job report lists the allocated payroll transactions by job. It shows the totals for the report dates and jobs you select for the report. Figure 17-13 shows an example of a memorized Labor Cost by Job report.

Memorizing the Labor Cost by Job Report — To memorize the Labor Cost by Job report in your QuickBooks Pro data file:

- From the **Reports** menu, choose **Accountant & Taxes**, and then **Transaction Detail by Account**.

- Select the columns in the Display tab of the Modify Report window, as shown in Figure 17-14. Select only **(left margin)**, **Type**, **Date**, **Source Name**, **Amount**, and **Balance**. Scroll down to the bottom of the list to find Amount and Balance.

- In **Report Date Range**, enter the date range you're interested in. Our example uses 12/7/2007 to 12/7/2007.

- In **Total by**, select **Account list** from the pull-down list.

- Click the **Filters** tab.

- Under **Filter**, select **Account**.

- From the pull-down Account list, select **Selected accounts** to bring up the pop-up window showing your Chart of Accounts, as shown in Figure 17-15.

- Select your Job Labor accounts. If you used our Chart of Accounts, the accounts are 5200, 5210, 5220, 5230, and 5240.

Chapter 17: Reports

Figure 17-14
Choose the columns to appear in the Labor Cost by Job report.

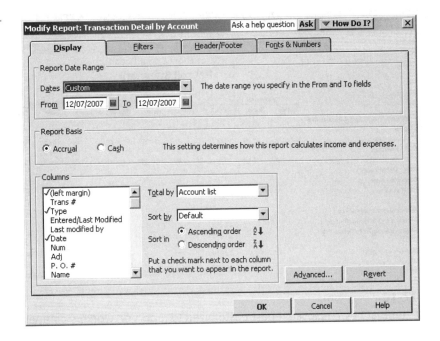

Figure 17-15
The Select Account list, showing our accounts for the Labor Cost by Job report.

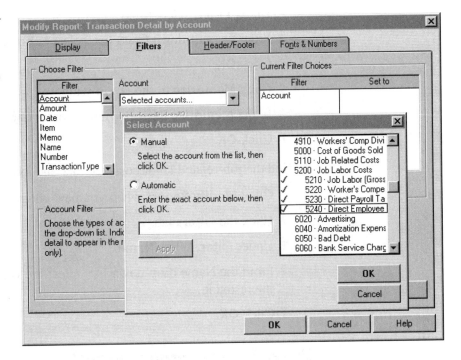

Contractor's Guide to QuickBooks Pro   **225**

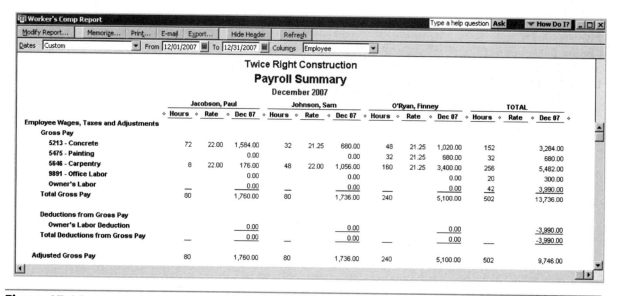

Figure 17-16
A Payroll Summary by Employee report, customized to report on workers' compensation.

- Click **OK** to close the Select Account window.
- Under **Filter**, select **Class**.
- From the pull-down Class list, select **Labor Costs - Job Related**.
- Click **OK**.
- Click **Memorize** in the report window.
- In the **Memorize Report** window, enter **Labor Costs by Job**.
- Click **OK**.

To find the job-related labor costs for just one job, do this:

- Click **Modify Report** when you have the report on your screen.
- Click the **Filters** tab.
- Under **Filter**, select **Name**.
- From the Name drop-down list, select the **customer:job** you're interested in.
- Click **OK**.

Workers' Compensation Report

The memorized Workers' Compensation Report lists the allocated payroll transactions by workers' compensation category. However, you must have used workers' comp codes as payroll items to use this report. The totals shown will be for the report dates you select. Figure 17-16 shows a memorized Workers' Comp (titled Payroll Summary) Report.

Figure 17-17
Set the dates for the Workers' Comp Report in this window.

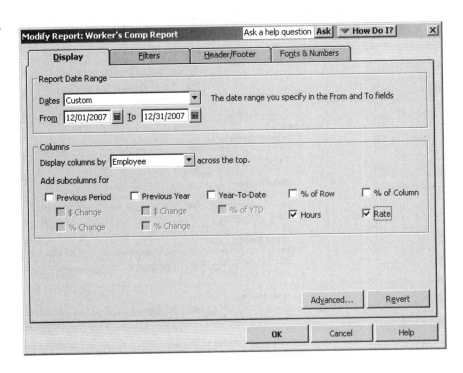

Memorizing the Workers' Compensation Report — To memorize the Workers' Compensation Report in your QuickBooks Pro data file:

- From the **Reports** menu, choose **Employees & Payroll**, and then **Payroll Summary**.
- On the Display tab of the Modify Report window, change the date range of the report to match the dates your Workers' Compensation Insurance carrier asks for, as shown in Figure 17-17.
- Click **OK**.
- Click **Memorize** in the report window.
- In the Memorize Report window, enter **Workers' Comp Report**.
- Click **OK**.

Job Costs Detail Report

The memorized Job Costs Detail report lists and totals all job cost transactions for any job you select, and summarizes the transactions by item or job phase. Figure 17-18 shows an example of a memorized Job Costs Detail report for December 1 through December 31, 2007. You'll need to adjust the dates of this report for your own business.

Chapter 17: Reports

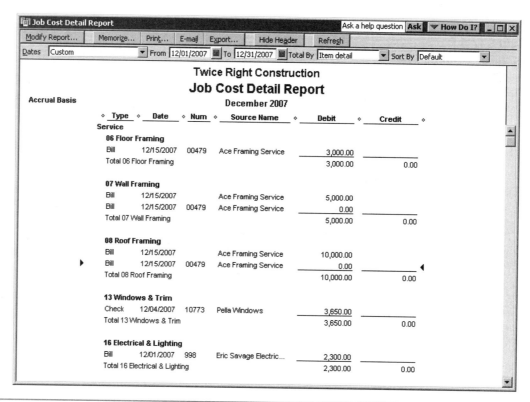

Figure 17-18
A memorized Job Costs Detail report.

Memorizing the Job Costs Detail Report — To memorize the Job Costs Detail report in your QuickBooks Pro data file:

- From the **Reports** menu, choose **Accountant & Taxes**, and then **Transaction Detail by Account**.
- Now select the columns in the Display tab of the Modify Report window, as shown in Figure 17-19. Select only **(left margin)**, **Type**, **Date**, **Num**, **Source Name**, **Debit**, and **Credit**. Scroll down to the bottom of the list to find Debit and Credit.
- From the Total by pull-down list, select **Item detail**.
- Click the **Filters** tab.
- Under **Filter**, select **Account**.
- From the pull-down Account list, select **All cost-of-sales accounts**.
- Under Filter, select **Transaction Type**.

228 *Contractor's Guide to QuickBooks Pro*

Figure 17-19
Choose the columns to appear in the Job Costs Detail report.

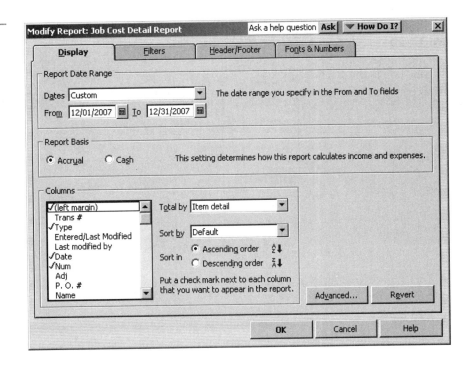

- From the pull-down Transaction Type, select **Selected Transaction Types** and then select **Check**, **Credit Card**, **Bill**, **CCard Refund**, **Bill Credit**, **Bill CCard**, **Item Receipt**, and **Paycheck**.
- Click **OK**.
- Under **Filter**, select **Date**. In **From**, enter the date you started entering transactions into QuickBooks Pro. In **To**, enter a date far into the future.
- Under **Filter**, select **Name**. In the **Name** drop-down list, select the job you wish to report on. See Figure 17-20.
- Click **OK**.
- Click **Memorize** in the report window.
- In the Memorize Report window, enter **Job Costs Detail**.
- Click **OK**.

Unpaid Job Bills

This report will list and total all unpaid job expenses. Transactions are grouped and subtotaled by vendor, which are the bold headings on the report. Figure 17-21 shows an example of a memorized Unpaid Job Bills report.

Contractor's Guide to QuickBooks Pro **229**

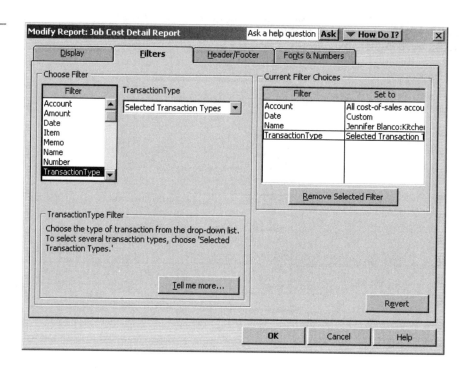

Figure 17-20
Choose the transaction types to include in the Job Costs Detail report.

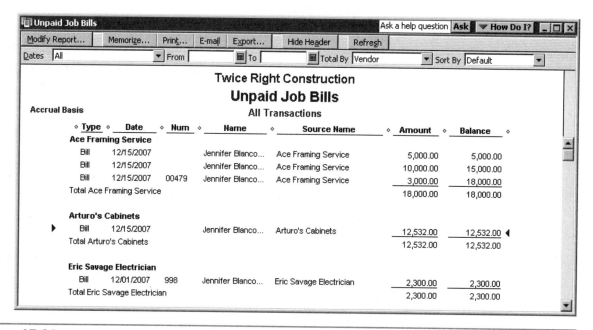

Figure 17-21
A memorized Unpaid Job Bills report.

Figure 17-22
Choose the columns to appear in the Unpaid Job Bills report.

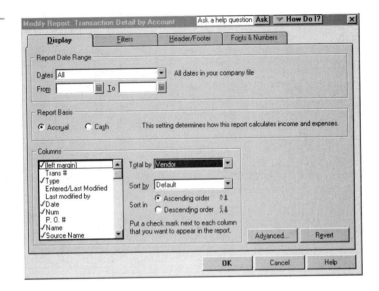

Memorizing the Unpaid Job Bills Report — To memorize the Unpaid Job Bills report in your QuickBooks Pro data file:

- From the **Reports** menu, choose **Accountant & Taxes**, and then **Transaction Detail by Account**.
- Now select the columns in the Display tab of the Modify Report window, as shown in Figure 17-22. Select only **(left margin)**, **Type**, **Date**, **Num**, **Name**, **Source Name**, **Amount**, and **Balance**. Scroll down to the bottom of the list to find Amount and Balance.
- In **Dates**, select **All**.
- In **Total by**, select **Vendor** from the pull-down list.
- Click the **Filters** tab.
- Under **Filter**, select **Account** and then from the pull-down Account list, select **All cost-of-sales accounts**. See Figure 17-23.
- Under **Filter**, select **Transaction Type** and then from the pull-down Transaction Type list, select **Bill**.
- Under **Filter**, select **Paid Status** and then **Open**, as shown in Figure 17-24.
- Click **OK**.
- Click **Memorize** in the report window.
- In the Memorize Report window, enter **Unpaid Job Bills**.
- Click **OK**.

Chapter 17: Reports

Figure 17-23
Choose the accounts to include in the Unpaid Job Bills report.

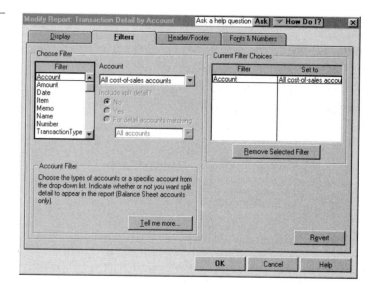

Figure 17-24
Choose the status of the transactions to include in the Unpaid Job Bills report.

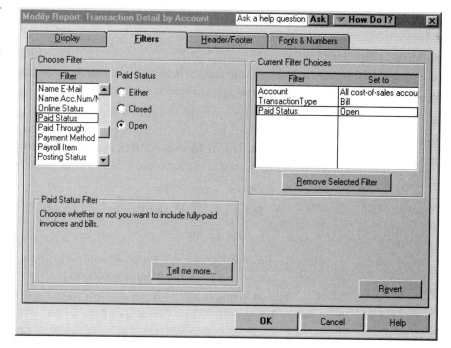

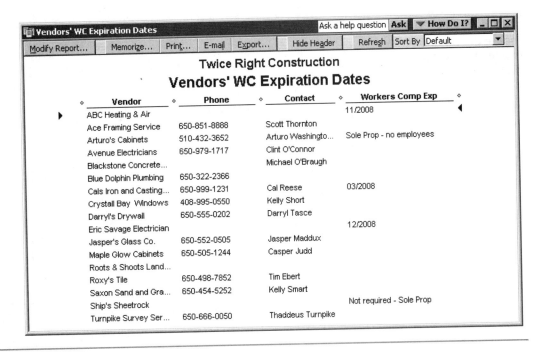

Figure 17-25
A memorized Vendors' WC Expiration Dates report.

Vendors' WC Expiration Dates

The memorized Vendor Workers' Comp Expiration Dates report lists all subcontractors, with the phone, contact name, and workers' comp expiration date. If you've made an entry into the custom field for "comp expires," that date will display in the far right column. Refer back to Chapter 14 for more information on creating and using the "comp expires" custom field. Figure 17-25 shows a memorized Vendors' WC Expiration Dates report.

Memorizing the Vendors' WC Expiration Dates Report — To memorize the Vendor's WC Expiration Dates report in your QuickBooks Pro data file:

- From the **Reports** menu, choose **Vendors & Payables** and **Vendor Contact List**.
- Now select the columns in the Display tab of the Modify Report window, as shown in Figure 17-26. Select only **(left margin)**, **Vendor**, **Phone**, **Contact**, and **Workers' Comp Expires**. We've scrolled down to the bottom of the list to show Workers' Comp Expires in the list, but it won't be there if you didn't set up a custom field for it.
- Click **OK**.
- Click **Memorize** in the report window.
- In the Memorize Report window, enter **Vendors' WC Expiration Dates**.
- Click **OK**.

Figure 17-26
Choose the columns to appear in the Vendors' WC Expiration Dates report.

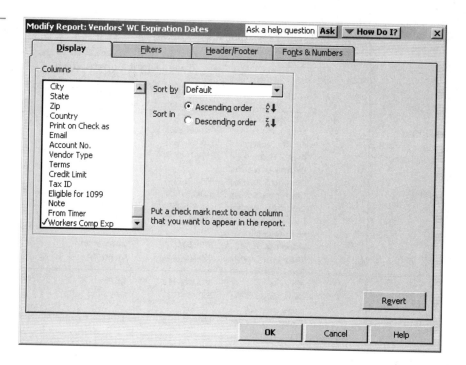

Using Jobs, Time & Mileage Reports

You can use the built-in QuickBooks Pro Jobs, Time & Mileage project reports to help you bill customers, analyze how well each job is progressing financially, and identify problem jobs before it's too late. We'll discuss what each report shows, how to use the report in your business, and when to run it. We'll also show you a sample printout of each report. We didn't discuss the Item Profitability, Item Estimates vs. Actuals, Profit & Loss by Job, or Time by Item reports because they aren't useful to contractors.

To run a project report:

> From the **Reports** menu, choose **Jobs, Time & Mileage** and select the report by name.

When the Modify Report window appears, you can usually accept the default values and just click OK to create the report. In some cases you'll want to specify a time period for the report.

When running certain reports, such as the Job Profitability Detail report, you'll be asked to select the job you want to run the report for in the Filter Report by Job window.

Figure 17-27
Use the Job Profitability Summary report to help you when you invoice customers.

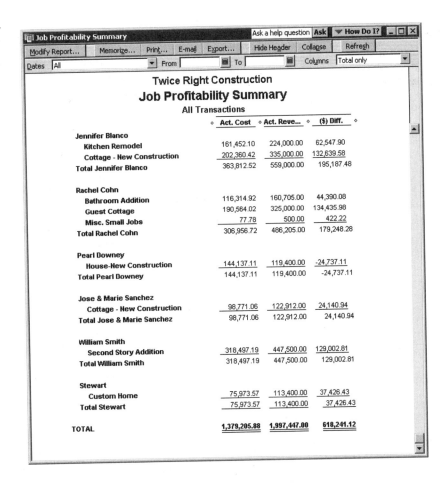

Job Profitability Summary Report

The Job Profitability Summary report lists every job you're working on for each of your customers. This report lists the jobs by name, actual job costs to date, actual revenue or total amount of billings to date, and in the last column, dollar difference. Dollar difference is the actual costs to date minus the total amount billed to the job in dollars. Figure 17-27 shows an example of a Job Profitability Summary.

Use this report to help you bill customers. Since the report gives you a good idea of where each project's actual costs are in relation to its billings, you can use it to get an idea of how much you should bill on each job. For example, if you bill your customers each week on a lump sum basis, this report will help you analyze exactly what the costs are to date and how much you've billed so far. Use this information to bill each job so your costs and profits are covered and the job is at or below its estimated cost.

If you've billed 100 percent of a job, the report will tell you how profitable it was. For example, if the actual cost of a job is $15,000 and you billed $17,000, the $2,000 difference is profit.

Print (and check) this report before you bill a customer.

Figure 17-28
The Job Profitability Detail report breaks down a particular job by item.

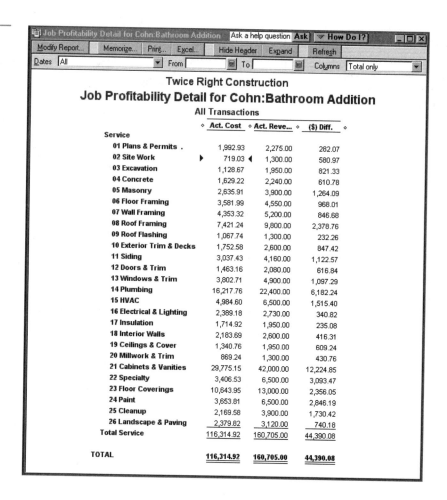

Job Profitability Detail Report

The Job Profitability Detail report shows a single job, broken down by job phases (items). It tells you the actual job costs to date, actual revenue or total amount of billings to date, and in the last column, the dollar difference. Dollar difference is the actual costs to date minus the total amount billed for each job phase, in dollars. Figure 17-28 shows a sample Job Profitability Detail report.

Use this report to help you bill a job on a percentage-complete basis by phase. Since this report gives you a good idea of where each project's actual costs are in relation to its billings, it'll help you decide how much you should bill by phase. For example, if you bill a job based on how much work has been completed on each phase of construction, use this report to analyze exactly what the costs are to date by job phase and how much you've previously billed. Then you can quickly produce a billing for the job by phase of construction.

Figure 17-29
Print this report often to check actual vs. estimated costs.

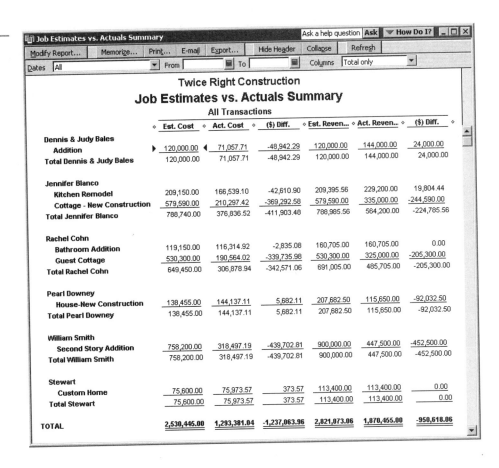

If you've billed 100 percent of a project, the report will give you an idea of each job's profitability by phase. For example, if phase 01 - Plans and Permits actual costs is $1,992.93 and its actual revenue is $2,275, then the profit is $282.07.

Print this report before you bill a customer. It'll help you make sure a job's actual costs aren't running too high or low in relation to its billings.

Job Estimates vs. Actuals Summary Report

The Job Estimates vs. Actuals Summary report tells you how each job is progressing, and which jobs are over, or under, their estimated costs. It's a handy report for quick reference, especially if you have many small jobs that last a short time. Print the report often to check actual versus estimated costs. Use this report to adjust estimated costs if you find that certain types of jobs consistently have lower profit margins. Figure 17-29 shows an example of a Job Estimates vs. Actuals Summary.

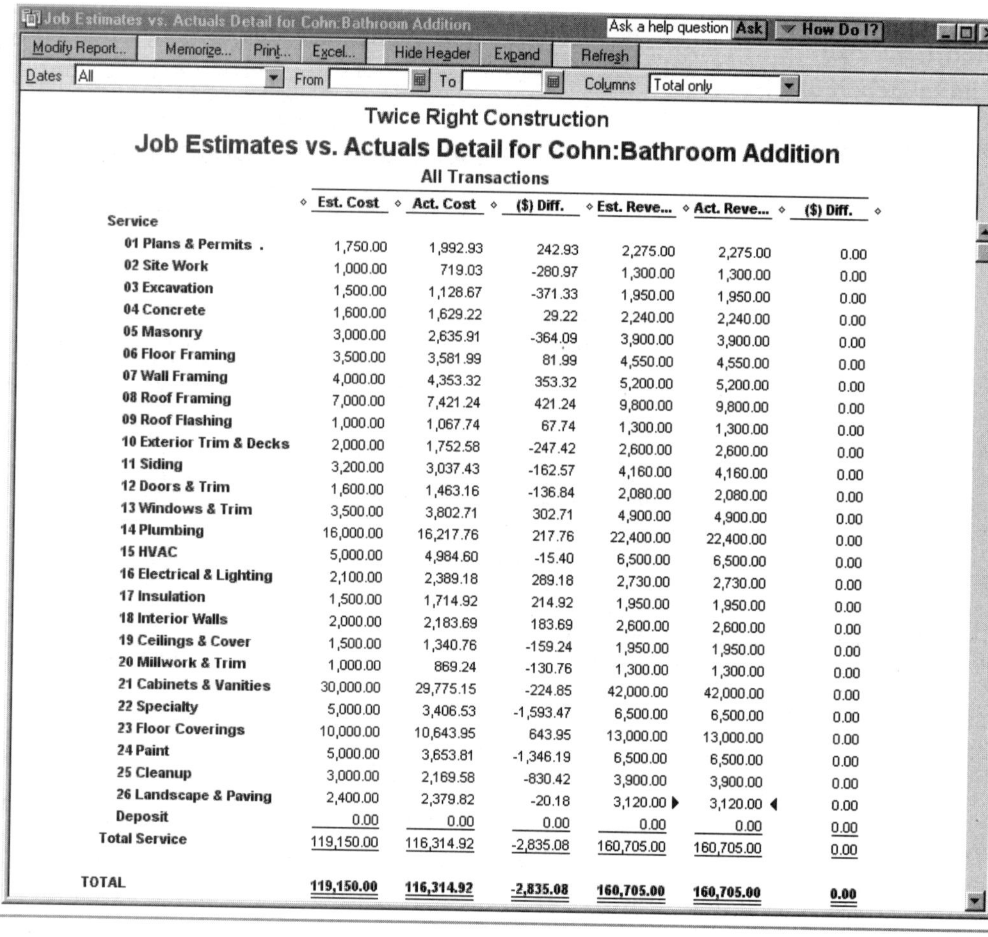

Twice Right Construction

Job Estimates vs. Actuals Detail for Cohn:Bathroom Addition

All Transactions

	Est. Cost	Act. Cost	($) Diff.	Est. Reve...	Act. Reve...	($) Diff.
Service						
01 Plans & Permits	1,750.00	1,992.93	242.93	2,275.00	2,275.00	0.00
02 Site Work	1,000.00	719.03	-280.97	1,300.00	1,300.00	0.00
03 Excavation	1,500.00	1,128.67	-371.33	1,950.00	1,950.00	0.00
04 Concrete	1,600.00	1,629.22	29.22	2,240.00	2,240.00	0.00
05 Masonry	3,000.00	2,635.91	-364.09	3,900.00	3,900.00	0.00
06 Floor Framing	3,500.00	3,581.99	81.99	4,550.00	4,550.00	0.00
07 Wall Framing	4,000.00	4,353.32	353.32	5,200.00	5,200.00	0.00
08 Roof Framing	7,000.00	7,421.24	421.24	9,800.00	9,800.00	0.00
09 Roof Flashing	1,000.00	1,067.74	67.74	1,300.00	1,300.00	0.00
10 Exterior Trim & Decks	2,000.00	1,752.58	-247.42	2,600.00	2,600.00	0.00
11 Siding	3,200.00	3,037.43	-162.57	4,160.00	4,160.00	0.00
12 Doors & Trim	1,600.00	1,463.16	-136.84	2,080.00	2,080.00	0.00
13 Windows & Trim	3,500.00	3,802.71	302.71	4,900.00	4,900.00	0.00
14 Plumbing	16,000.00	16,217.76	217.76	22,400.00	22,400.00	0.00
15 HVAC	5,000.00	4,984.60	-15.40	6,500.00	6,500.00	0.00
16 Electrical & Lighting	2,100.00	2,389.18	289.18	2,730.00	2,730.00	0.00
17 Insulation	1,500.00	1,714.92	214.92	1,950.00	1,950.00	0.00
18 Interior Walls	2,000.00	2,183.69	183.69	2,600.00	2,600.00	0.00
19 Ceilings & Cover	1,500.00	1,340.76	-159.24	1,950.00	1,950.00	0.00
20 Millwork & Trim	1,000.00	869.24	-130.76	1,300.00	1,300.00	0.00
21 Cabinets & Vanities	30,000.00	29,775.15	-224.85	42,000.00	42,000.00	0.00
22 Specialty	5,000.00	3,406.53	-1,593.47	6,500.00	6,500.00	0.00
23 Floor Coverings	10,000.00	10,643.95	643.95	13,000.00	13,000.00	0.00
24 Paint	5,000.00	3,653.81	-1,346.19	6,500.00	6,500.00	0.00
25 Cleanup	3,000.00	2,169.58	-830.42	3,900.00	3,900.00	0.00
26 Landscape & Paving	2,400.00	2,379.82	-20.18	3,120.00	3,120.00	0.00
Deposit	0.00	0.00	0.00	0.00	0.00	0.00
Total Service	119,150.00	116,314.92	-2,835.08	160,705.00	160,705.00	0.00
TOTAL	**119,150.00**	**116,314.92**	**-2,835.08**	**160,705.00**	**160,705.00**	**0.00**

Figure 17-30

The Job Estimates vs. Actuals Detail report breaks down a particular job by item.

If you tend to have larger projects with many phases that span several months or years, this report may not be as useful to you as the next one, Job Estimates vs. Actuals Detail report.

Job Estimates vs. Actuals Detail Report

The Job Estimates vs. Actuals Detail report tells you, by job phase, how each job is progressing, and which part of a job is in danger of running over its estimated costs. Figure 17-30 shows a sample Job Estimates vs. Actuals Detail report.

This is a handy report for quick reference, especially if you have jobs that last a long time and you need to track them by job phase. You can also use this report to help estimate future jobs. It'll give you a good idea of which phases were over, or under, estimate.

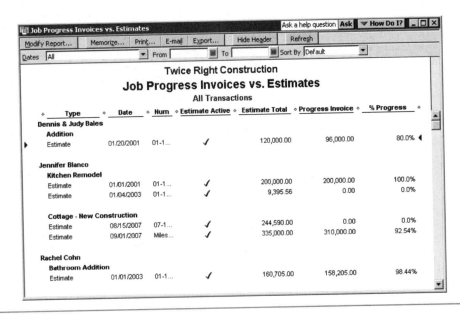

Figure 17-31
Use the Job Progress Invoices vs. Estimates report to check that the job site progress matches the percentage of progress.

Print this report once a week and give it to a job's Project Manager. Then he can see which part or parts of his project are over, or under, estimate while there's still time to make adjustments.

Job Progress Invoices vs. Estimates Report

The Job Progress Invoices vs. Estimates report lists each customer:job, job status, estimate total, amount invoiced to date, and percentage complete. Figure 17-31 shows a sample Job Progress Invoices vs. Estimates report.

This report gives you an idea of how much has been billed to date (in dollars) and what percentage that amount is of the total estimated costs. It's important to run this report from time to time to check whether the actual job site progress matches up with the percentage progress. For example, if you get a project status report from the field each week, compare the estimated field percentage complete to the % Progress on this report. They should be similar. If the information from the field estimates a higher percentage complete than the % Progress report, you may be behind in your billings to the customer. If the % Progress report has a higher percentage complete than the actual field progress estimate, you may have billed more than you should have in the last billing cycle.

Chapter 17: Reports

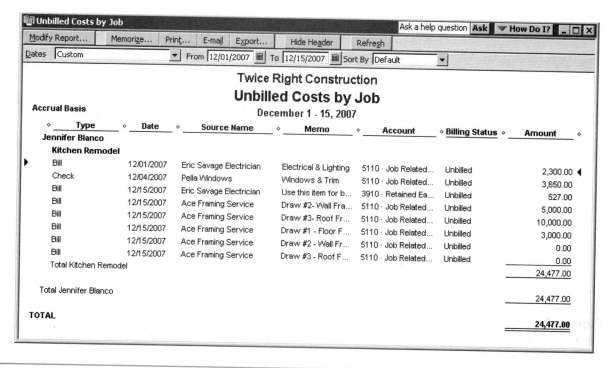

Figure 17-32
An Unbilled Costs by Job report.

You may need to adjust job status as work progresses. To do this:

❚ From the **Lists** menu, choose **Customer:Job List**.

❚ Double-click the job.

❚ Click the **Job Info** tab.

❚ In **Job Status**, select the new status from the pull-down list.

❚ Click **OK**.

Unbilled Costs by Job Report

This report is handy if you bill your customers on a time-and-materials basis. It gives you a list of all project costs (except labor costs) that haven't been billed to a job. Use this report to double-check that you've billed all costs on a time-and-materials basis to a job. Figure 17-32 shows an example of an Unbilled Costs by Job report.

240 Contractor's Guide to QuickBooks Pro

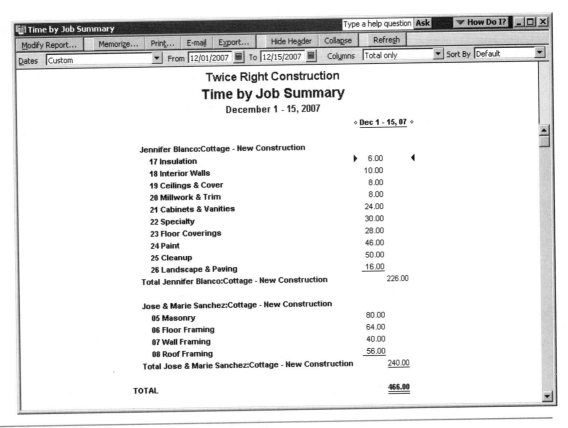

Figure 17-33
The Time by Job Summary report.

Time by Job Summary Report

This report lists each job, with detailed information for the job and phase. It's particularly useful if a job seems to be running over budget and you suspect that time has been mistakenly allocated to the wrong job or phase. Figure 17-33 shows an example of a Time by Job Summary report.

Open Purchase Orders by Job Report

In Chapter 14 we described how to use purchase orders to track committed costs or subcontractors' bids. This report will help you figure out how much you still have outstanding on those committed costs or subcontractors' bids.

Chapter 17: Reports

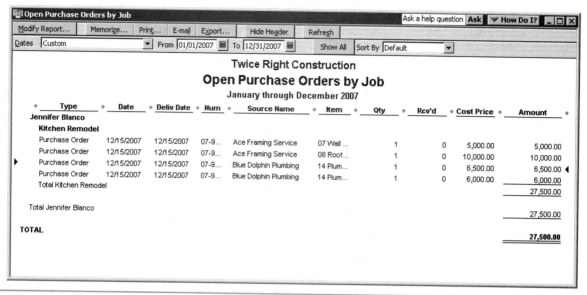

Figure 17-34
An Open Purchase Orders by Job report.

QuickBooks Pro has a built-in report for tracking purchase orders that haven't been filled (in other words, a report listing all subcontractors who haven't completed or submitted bills on a project). Figure 17-34 shows an example of an Open Purchase Orders by Job report.

Important! It's worth noting here that you may have a problem with this report if you make partial payments to subcontractors. The problem occurs when you release a portion of the original PO. The remaining amount will clear from the report if you don't enter a quantity *less than 1* in the quantity field. In other words, to avoid reporting problems you should always enter an amount less than 1 when you're releasing only a portion of the PO. For more information, see Chapter 14 under the heading Using Purchase Orders to Track Multiple Draws and Committed Costs Buyouts.

The Delivery Date represents the day you were told the goods would arrive (for example, special order windows or doors) or the date the services were to start (the date the plumber or electrician said he would start work).

Time by Job Detail Report

This report lists each employee and his/her time on each job phase, subtotaled by job. Use this report to double-check that the time was entered correctly for each timecard, to calculate the total hours spent on a job, or to see who's spending time on specific jobs. Figure 17-35 shows an example of a Time by Job Detail report.

Figure 17-35
A Time by Job Detail report.

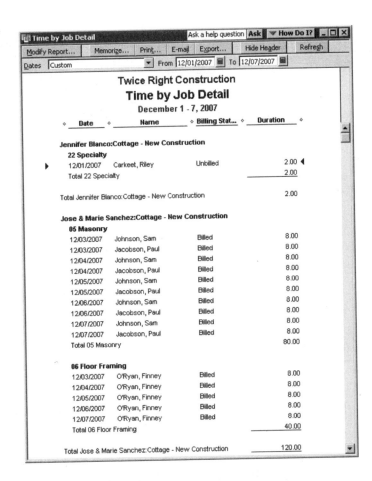

Time by Name Report

This report lists all employees and the hours each worked during a certain period. Use this report to find out how many hours each person has worked, and on which job, during a time period. Figure 17-36 shows part of a sample Time by Name report.

In this chapter we showed you how to modify, filter, and memorize reports. We also explained the built-in QuickBooks Pro project reports.

Along with these reports, you also need to set up procedures to check your data entry. As an example, you need to be sure all your customers are being invoiced and that your Accounts Receivable are in balance. The same applies to your bills and Accounts Payable. These procedures can be broken down into end-of-month procedures and end-of-year procedures. The next chapter will walk you through those procedures, step-by-step.

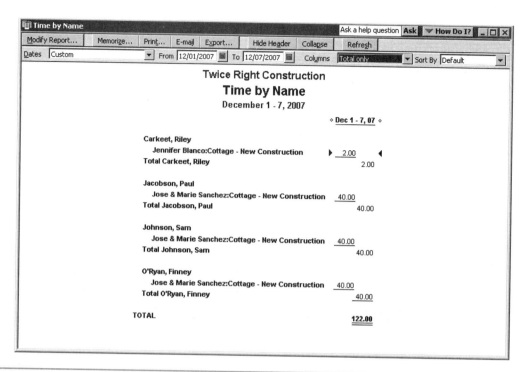

Figure 17-36
A Time by Name report.

Using Reports in QuickBooks Premier: Contractor Edition

Some valuable reports are included in the Contractor Edition of QuickBooks. We've outlined two important reports below.

To run these reports:

▌ From the **Reports** menu, choose **Contractor Reports** and select the report by name.

Cost to Complete by Job Detail

The Cost to Complete by Job Detail report is excellent for giving the contractor an idea of where they stand on any given job, based on the percentage complete by phase of the project.

When running this report, you will be prompted to filter the report. Select the appropriate Customer:Job and click **OK**.

Enter the percent complete for each job phase in the % Complete column. See Figure 17-37. Click on **OK**.

In the Modify Report window, click on **OK** to display the report. See Figure 17-38.

Figure 17-37
Enter the percent complete for each job phase.

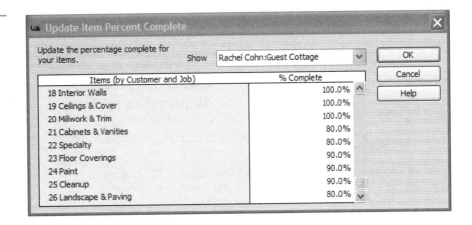

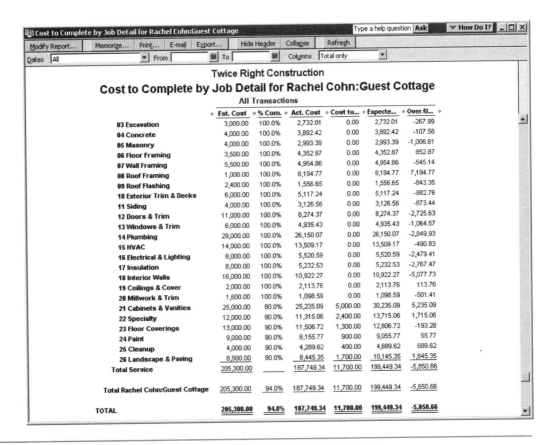

Figure 17-38
A Cost to Complete by Job Detail report.

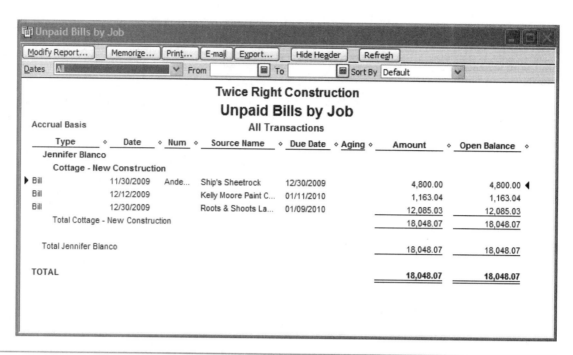

Figure 17-39
An Unpaid Bills by Job report.

Unpaid Bills by Job

The Unpaid Bills by Job report sorts all unpaid bills by job. If you pay your vendors when you receive a check from a customer and need to know which bills are outstanding for a job, you would print this report. See Figure 17-39.

Chapter 18
End of Month and End of Year Procedures

In this chapter, we'll explain what you should do at the end of each month and at the end of each fiscal year. Keep in mind that QuickBooks Pro doesn't force you to close out any month or year. We're simply suggesting what you should do to make sure that the information you've entered into QuickBooks Pro is accurate. We also suggest that you print out several reports to help you analyze how well your business is doing financially.

End of Month Procedures

Let's look first at what to do at the end of each month. You should:

- Reconcile your checking account with your bank statement.
- Run reports to make sure all transactions are correct.
- Print monthly reports.
- Back up data and close the month.

Reconcile Each Checking Account

QuickBooks Pro makes bank account reconciliation simple, so you won't have any excuse for putting it off. Set aside the time and do it every month. Reconciling your bank account will catch any errors you or your bank have made, and serve as a check of your own accounting. Here's how to do it:

- From the **Banking** menu, choose **Reconcile**.
- At the Begin Reconciliation window, in **Account**, select your checking account from the pull-down list. In our example shown in Figure 18-1, it's 1110 Company Checking Account.
- In **Statement Date**, enter the ending date on the bank statement.
- In **Ending Balance**, enter the amount on the bank statement (usually called New Balance or Balance this Statement).

Chapter 18: End of Month and End of Year Procedures

Figure 18-1
Use the Begin Reconciliation window to select your checking account and enter all relevant information.

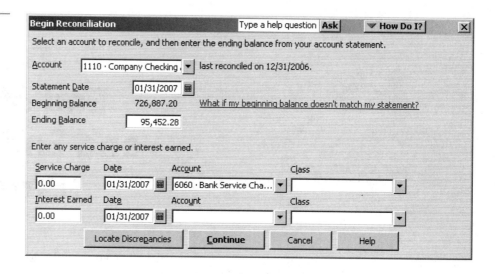

- In **Service Charge**, enter the amount listed on the statement, and the date the service charge was posted. In **Account**, select the expense account Bank Service Charges in your Chart of Accounts from the pull-down list.

- In **Interest Earned**, enter the amount listed on the statement, and the date any interest was posted. In **Account**, select the income account Interest Income in your Chart of Accounts from the pull-down list.

- Click **Continue**.

At the Reconcile window (Figure 18-2), verify each deposit, credit, check, payment, and service charge to make sure it's the same as the corresponding amount on the bank statement. If an amount is different, you have to find the problem and correct it before continuing. To make a correction, double-click the transaction and make the appropriate changes. If the bank made an error, make sure they correct it. If you find an amount on the bank statement that isn't in your company data file, it may mean you forgot to enter it. Enter it now before continuing with the reconciliation.

When your records match the bank statement, Ending Balance and Cleared Balance will be the same, and Difference will be zero.

- Click **Reconcile Now** to complete the reconciliation.

Now you'll get a dialog box asking what type of reconciliation report you would like to print:

- Select **Detail**.
- Click **Print**.

Figure 18-2
Use the Reconcile window to clear checks and deposits according to your bank statement.

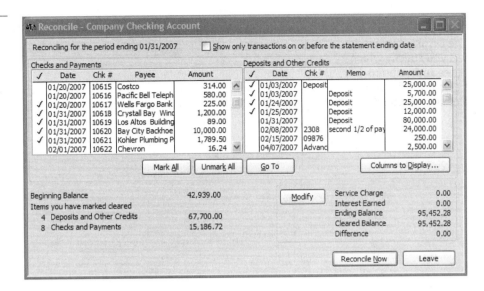

We suggest you file the printed reconciliation report and bank statement together, or set up a three-ring binder to store the reports.

Run Reports to Make Sure All Transactions Are Correct

The three reports you should run to make sure you've entered all transactions correctly are:

- Transaction Detail by Account
- A/R Aging Detail
- A/P Aging Detail

On each report, check each transaction to make sure it's correct and posted to the correct account. You have to correct any mistakes before continuing. While you have each report window open on your computer, you can reopen any transaction by double-clicking it. When you do that, you'll get a window showing the original transaction. You can make any corrections you need in that window.

Now let's look at each one of these reports.

Transaction Detail by Account — The Transaction Detail by Account report lists each account in your Chart of Accounts and all transactions on each account. You can use this report to find any transactions that have been posted to the wrong account. For example, all checks written to the phone

Figure 18-3

The Display tab window lets you specify which dates and columns will be on the Transaction Detail by Account report.

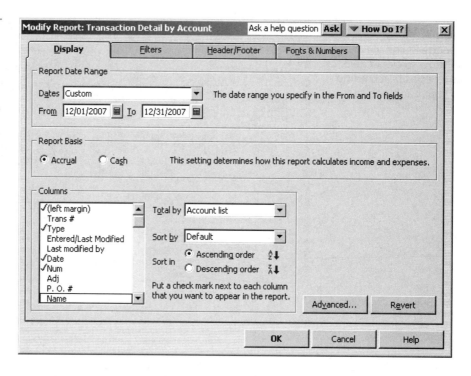

company should be posted to the Telephone account. If a check for office supplies shows up under the Telephone account, you'll need to correct the transaction. Here's how to get the Transaction Detail by Account report:

- From the **Reports** menu, choose **Accountant & Taxes**, and then **Transaction Detail by Account**. Select the full month you're closing. In the examples we show here, we've used the month of December 2007.

- In the Display tab of the Modify Report window, under the Columns box, select only **(left margin)**, **Type**, **Date**, **Num**, **Source Name**, **Memo**, **Class** (optional), **Amount**, and **Balance** as shown in Figure 18-3. You'll have to scroll down to the bottom of the list to find Memo, Class, Amount, and Balance.

- In **Dates**, select **Custom**. In **From** and **To**, enter the month you're closing. Ignore the remaining choices.

- Click **OK**.

Figure 18-4 shows an example of this customized Transaction Detail by Account report. Check the transactions on your own report and correct any that are in error.

A/R Aging Detail Report — The A/R Aging Detail report lists all unpaid customer invoices as of the last day of the month you're closing. You can use it to show any invoices that have been posted twice, missing invoices, and payments that haven't been applied to an invoice. You'll know there's a mistake if you see a payment with a negative amount. To fix this problem,

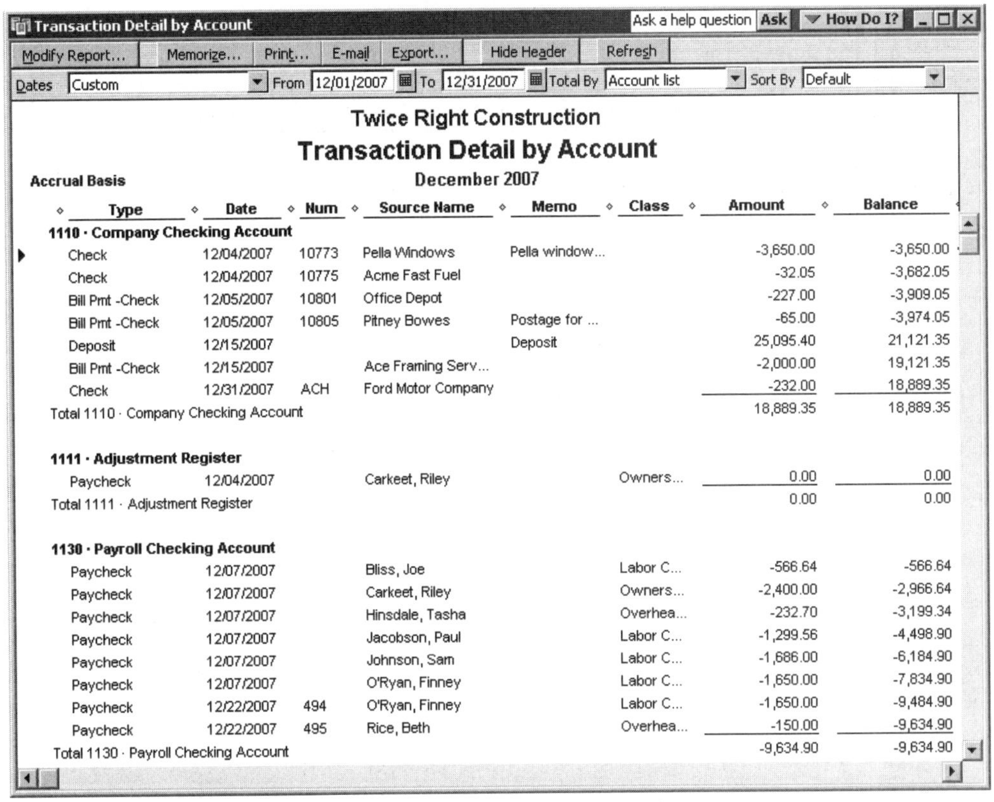
Transaction Detail by Account

Modify Report... | Memorize... | Print... | E-mail | Export... | Hide Header | Refresh

Dates Custom | From 12/01/2007 | To 12/31/2007 | Total By Account list | Sort By Default

Twice Right Construction

Transaction Detail by Account

Accrual Basis — December 2007

Type	Date	Num	Source Name	Memo	Class	Amount	Balance
1110 · Company Checking Account							
Check	12/04/2007	10773	Pella Windows	Pella window...		-3,650.00	-3,650.00
Check	12/04/2007	10775	Acme Fast Fuel			-32.05	-3,682.05
Bill Pmt -Check	12/05/2007	10801	Office Depot			-227.00	-3,909.05
Bill Pmt -Check	12/05/2007	10805	Pitney Bowes	Postage for ...		-65.00	-3,974.05
Deposit	12/15/2007			Deposit		25,095.40	21,121.35
Bill Pmt -Check	12/15/2007		Ace Framing Serv...			-2,000.00	19,121.35
Check	12/31/2007	ACH	Ford Motor Company			-232.00	18,889.35
Total 1110 · Company Checking Account						18,889.35	18,889.35
1111 · Adjustment Register							
Paycheck	12/04/2007		Carkeet, Riley		Owners...	0.00	0.00
Total 1111 · Adjustment Register						0.00	0.00
1130 · Payroll Checking Account							
Paycheck	12/07/2007		Bliss, Joe		Labor C...	-566.64	-566.64
Paycheck	12/07/2007		Carkeet, Riley		Owners...	-2,400.00	-2,966.64
Paycheck	12/07/2007		Hinsdale, Tasha		Overhea...	-232.70	-3,199.34
Paycheck	12/07/2007		Jacobson, Paul		Labor C...	-1,299.56	-4,498.90
Paycheck	12/07/2007		Johnson, Sam		Labor C...	-1,686.00	-6,184.90
Paycheck	12/07/2007		O'Ryan, Finney		Labor C...	-1,650.00	-7,834.90
Paycheck	12/22/2007	494	O'Ryan, Finney		Labor C...	-1,650.00	-9,484.90
Paycheck	12/22/2007	495	Rice, Beth		Overhea...	-150.00	-9,634.90
Total 1130 · Payroll Checking Account						-9,634.90	-9,634.90

Figure 18-4

A sample Transaction Detail by Account report you can create in QuickBooks Pro.

double-click the negative payment amount, and at the bottom of the Customer Payment window, select the invoice this payment should be applied to.

Here's how to get an A/R Aging Detail report like the example we show in Figure 18-5:

- From the **Reports** menu, choose **Customers & Receivables**, and then **A/R Aging Detail**.
- In the Display tab of the Modify Report window, under the Columns box, select only **(left margin)**, **Type**, **Date**, **Num**, **Source Name**, **Due Date**, **Aging**, and **Open Balance** as shown in Figure 18-6. You'll have to scroll down to the bottom of the list to find Due Date, Aging and Open Balance.
- In **Dates**, select **Custom**. In **From** and **To**, enter the month you're closing. Ignore the remaining choices.
- Click **OK**.

Check the transactions on your own report and correct any errors.

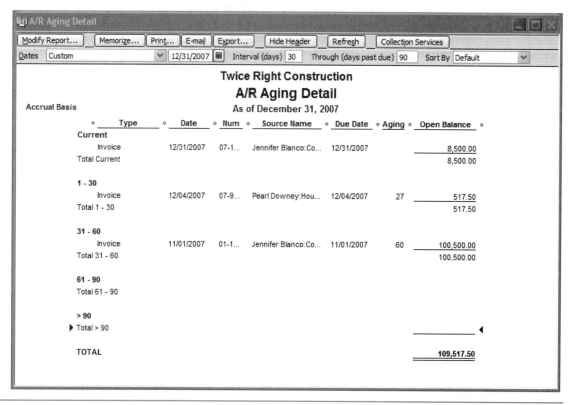

Figure 18-5
A sample A/R Aging Detail report you can create in QuickBooks Pro.

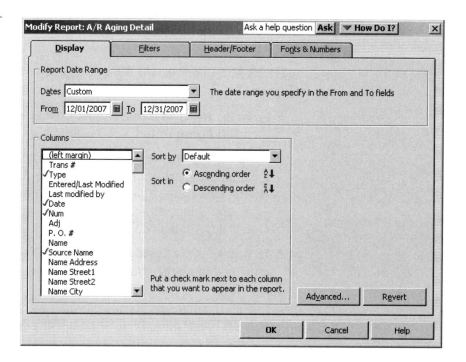

Figure 18-6
The Display tab window lets you specify which dates and columns will be on the A/R Aging Detail report.

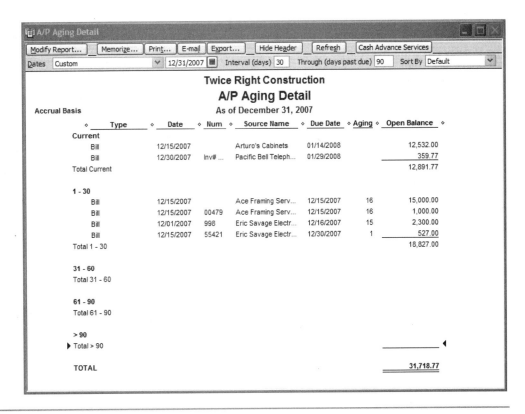

Figure 18-7
A sample A/P Aging Detail report you can create in QuickBooks Pro.

A/P Aging Detail Report — The A/P Aging Detail report lists all unpaid vendor bills as of the last day of the month you're closing. You can use it to show any missing bills or bills that have been posted twice.

Here's how to get an A/P Aging Detail report like the one we show in our example in Figure 18-7:

▪ From the **Reports** menu, choose **Vendors & Payables**, and then **A/P Aging Detail**.

▪ In the Display tab of the Modify Report window, under the Columns box, select only **(left margin)**, **Type**, **Date**, **Num**, **Source Name**, **Due Date**, **Aging**, and **Open Balance** as shown in Figure 18-8. You'll have to scroll down to the bottom of the list to find Due Date, Aging and Open Balance.

▪ In **Dates**, select **Custom**. In **From** and **To**, enter the month you're closing. Ignore the remaining choices.

▪ Click **OK**.

Check all the transactions on your own report and correct as necessary. Be sure to watch out for missing bills and bills that have been posted twice.

Figure 18-8
The Display tab window lets you specify which dates and columns will be on the A/P Aging Detail report.

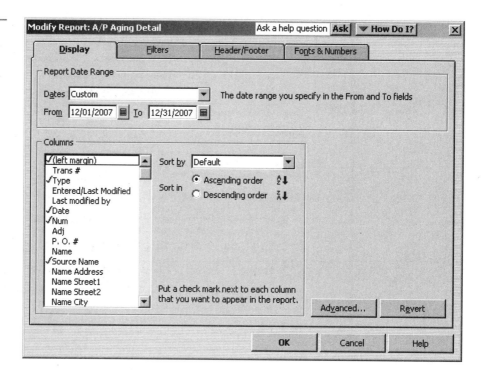

Printing and Filing Monthly Reports

Now you're ready to print these reports for the month you're closing. To print the Transaction Detail by Account and the A/R and A/P Aging Detail reports, just follow the same steps you used to open and correct the reports, and then click Print in each report window.

You'll also want to print financial statements for the month. To do that:

- From the **Reports** menu, choose **Company & Financial**, and then **Balance Sheet Standard**.
- Click **OK** to close the Modify Report window, then click **Print**.
- From the **Reports** menu, choose **Company & Financial**, and then **Profit & Loss YTD Comparison**.
- Click **OK** to close the Modify Report window, then click **Print**.

Back Up Data and Close the Month

When you've made sure everything is correct, you're ready to back up your data. To do that:

- From the **File** menu, choose **Back Up**.

- In the QuickBooks Backup window, on the Back Up Company File tab, you'll be asked to enter a name and location for your backup file. The name will automatically default to the name of your company file, except it will have the extension QBB instead of QBW. In the location field, select a floppy or zip drive to store the backup copy on.

- Click **OK** when you've selected the appropriate location for the backup file.

Now you're ready to close and lock yourself out of making changes to the accounting period you just completed.

- From the **Edit** menu, choose **Preferences**.

- Click the **Accounting Preferences** icon in the left hand scroll bar.

- In the Closing Date area, enter the last day of the month just closed. This will keep you from making any entries into the month you just closed.

- Click **OK**.

Reading and Understanding Your Financial Reports

It's important to take some time to analyze how well your business is doing before you file your reports. The rest of the work up to this point was just that, work. Now we get to have a little fun. Let's analyze the past to change the future. You can pinpoint your overexpenditures and figure out, finally, what your overhead percentage is running.

Balance Sheet

The Balance Sheet lists:

- your company's assets (things your company owns).
- liabilities (things your company owes money for).
- capital or equity (assets minus the liabilities or the amount of equity you've built up in the business).

There are two things to look for in this report. First, make sure everything is listed on the report. Are your checking account, savings account, company vehicles, large equipment, and so on, listed as assets? Then check the liabilities. Are all loans, lines of credit, and payroll taxes you owe listed?

Second, let's analyze the report. Keep in mind that this is the report the bank will look at to determine how well you manage your business, as well as its stability and growth. The bank will be looking closely at the assets in relation to the liabilities. The bank wants to know if your company is solvent. In other words, if you stopped business today, could your company sell its assets and pay off all its liabilities and have a few bucks left over? The bank wants to see total assets of two times your total liabilities, give or take a few dollars.

Balance Sheet Comparison (Optional Report for Analysis)

It's best if you analyze the Balance Sheet over two periods — for example first quarter 2007 and second quarter 2007 (or better yet, year to year).

If you've been using QuickBooks Pro for more than one quarter, print the Balance Sheet Comparison report to see how you did this quarter compared to last quarter. To print this report:

- From the **Reports** menu, choose **Company & Financial**, and then **Balance Sheet Prev Year Comparison**.
- Click **OK** to close the Modify Report window, then click **Print**.

On this report, check to see:

- Are the assets increasing?
- Are the liabilities decreasing?
- If the liabilities increased, the assets should have increased as well.
- Is the value of the business (assets minus liabilities) increasing or decreasing?

If it's decreasing, something is wrong. Give the problem your immediate attention.

Profit & Loss YTD Comparison Report

The Profit & Loss YTD Comparison report shows how much income your company has generated versus its expenses for the same time period. It's obvious here that you want to be generating more income than expenses, and the bottom line or net income/loss needs to be positive. This report will also list your total job costs and your overhead percentage. In Figure 18-9, we show an example of the report. We've modified the report to include the % of Income column.

Take a look at the Total Cost of Goods Sold (COGS) row and check the percentage in the column called % YTD. In the construction business, this percentage averages about 40 to 60 percent. If your percentage is lower, good for you! That means more money is going in your pocket. You're

Chapter 18: End of Month and End of Year Procedures

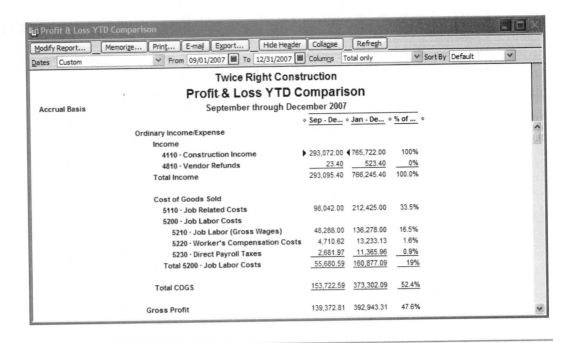

Figure 18-9
A sample Profit & Loss YTD Comparison report for the last quarter of 2007.

spending less on construction-related costs than the average construction company. In Figure 18-9, it's 52.4%. That's right where we want to be.

Now scroll down to the Total Expense row and check the percentage in the column called % YTD. This is the percentage of your income you spend for overhead costs — costs that you can't associate with a job expense, such as office rent, telephone, postage, and utilities. If your percentage is 10, it means that for every $100.00 you generate in revenue, you spend $10.00 for overhead costs. In the construction business, the average is about 8 to 12 percent.

To print this report:

▌ From the **Reports** menu, choose **Company & Financial**, and then **Profit & Loss YTD Comparison**.

▌ Click **OK** to close the Modify Report window, then click **Print**.

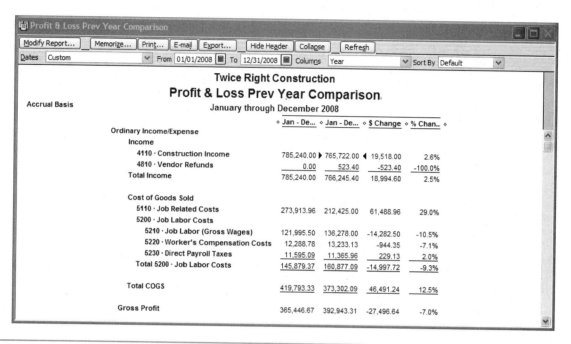

Figure 18-10
A sample Profit & Loss Previous Year Comparison report comparing 2007 with 2008.

Profit & Loss Prev Year Comparison Report

Again, like the Balance Sheet, it's best if you analyze the Profit & Loss over two years. If you've been using QuickBooks Pro for more than one year, print the Profit & Loss Prev Year Comparison report to see how you did this year compared to last year. In Figure 18-10, we show an example of a report comparing 2007 with 2008.

This report will tell you how well you did last year compared to this year. First, compare your Total Income. Has it increased or decreased from last year? Second, compare your Total COGS — where COGS = Job Costs. Have your job-related costs gone up or down? Why would that be? Third, look at Total Expense. Have your expenses gone up or down? Are you spending more money this year than last? Where are you spending less?

File the monthly reports in the company accounting binder for the appropriate year. Label the tabs in the binder according to the names of the reports you printed — Balance Sheet and Profit & Loss, for example.

To print this report:

■ From the **Reports** menu, choose **Company & Financial**, and then **Profit & Loss Prev Year Comparison**.

■ Click **OK** to close the Modify Report window, then click **Print**.

End of Year Procedures

In QuickBooks Pro you don't have to end the current year before you start a new year. To end your year:

- Follow the procedures in the previous section to close each month.
- Write off bad debts.
- Print your financial reports, file them, and give a copy to your tax preparer.
- Back up your QuickBooks file and close and lock the year.

Write Off Bad Debts and Run the Aging Detail Reports

At the end of the year you'll want to write off any bad debts from customers you know aren't going to pay you. To do this, create a credit memo for each customer invoice that won't be paid. This will allow you to clear the invoice out of Accounts Receivable and decrease your net profit by the amount of the unpaid invoice.

Make sure every invoice listed was actually outstanding as of the last day of your fiscal year. Check that you've entered all customer payments. Then run the A/R Aging Detail report, using the last day of your fiscal year as both the To and From dates.

Print Your Annual Reports

At the close of your fiscal year, print the Balance Sheet, Profit & Loss statement, and aging reports for the year you're closing.

From the **Reports** menu, choose and print these reports:

- Company & Financial, Balance Sheet Standard
- Company & Financial, Profit & Loss Standard
- Customers & Receivables, A/R Aging Detail
- Vendors & Payables, A/P Aging Detail

File the annual reports in the company accounting binder for the appropriate year. Label the tabs in the binder according to the names of the reports you printed — Balance Sheet, Profit & Loss, for example.

Make a copy of the Balance Sheet and Profit & Loss reports for your tax preparer.

Back Up Your Data Files and Close the Year

To back up your end of the year data:

- From the **File** menu, choose **Back Up.**

- In the QuickBooks Backup Window, on the Back Up Company File tab, you'll be asked to enter a name and location for your backup file. The name will automatically default to the name of your company file except it will have the extension QBB instead of QBW. In the location field, select a floppy or zip drive to store the backup copy on.

- Click **OK** when you have selected the appropriate location for the backup file.

Now you're ready to close and lock yourself out of making changes to the period you just completed.

- From the **Edit** menu, choose **Preferences**.

- Click the **Accounting Preferences** icon in the left hand scroll bar.

- In the Closing Date area, enter the last day of the year just closed. This will keep you from making any entries into the year you just closed.

- Click **OK**.

In this chapter, you learned what we recommend you do at month end and at year end. You saw how to reconcile your bank statements, print financial reports, and review transactions to make sure they were entered correctly. The next chapter is for builders who develop land. If you're a builder, be sure to go through this chapter. If you're a remodeling contractor and you don't develop land, nor plan to, you may want to just skim over the next chapter. But read the conclusion. It's pertinent to all contractors.

Chapter 19

Real Estate Development

As a spec builder or developer, you have to make sure your accounting system handles some specialized tasks. You need to track the money you borrow and spend to purchase land, improve that land, and then build on the improved land. Then you also need to record the sale of the property. In this chapter, we'll look at how you can use QuickBooks Pro to easily deal with these matters.

New Accounts

First, you may need to add a few new accounts to your Chart of Accounts. If you need more information on setting up accounts, see Chapter 3. Here are the new accounts you'll need:

Account number	Name	Type
1111	Adjustment Register	Bank
1460	Escrow Deposit	Other Current Asset
1470	Land Purchase	Other Current Asset
1471	Land Interest/Closing Costs	Other Current Asset
1480	WIP — Land Development	Other Current Asset
1490	WIP — Construction	Other Current Asset
2300	Loans Payable	Other Current Liability
2400	Land Acquisition Loan	Other Current Liability
2405	Land Development Loan	Other Current Liability
2410	Construction Loan	Other Current Liability

▮ *Adjustment Register* — Use the Adjustment Register account to temporarily record the transaction where you pay off one loan with another loan. For example, a development loan may pay off a land loan, or a construction loan may pay off a development loan. You'll see later exactly how to use this account to temporarily record loans that have been paid off during escrow closing. In our example Chart of Accounts, we use 1111 Adjustment Register.

- *Escrow Deposit Account* — Use the Escrow Deposit account if you're required to place a deposit with an escrow company when you purchase land. Usually you write a check to the escrow company to open up an escrow account. In our example Chart of Accounts, we use 1460 Escrow Deposit.

- *Land Purchase Accounts* — Use the Land Purchase accounts to keep track of the original purchase price of any land you purchase for development and costs related to that purchase. In our example Chart of Accounts, we use 1470 Land Purchase and 1471 Land Interest/Closing Costs.

- *WIP — Work in Process Accounts* — Use the WIP accounts to keep track of the costs of developing and building on a property. These accounts will appear on your balance sheet as assets, even though you use them for costs. In real estate development, you have to hold your costs until you sell the property. Then you match the sale of the building and the construction costs in the same time period. Once you sell the building, you'll transfer the WIP totals from the WIP accounts to the Job Related Costs account. In our example Chart of Accounts, we use 1480 WIP Land Development and 1490 WIP Construction.

- *Loan Accounts* — The loan accounts should total the amount of funds you borrowed for land acquisition, land development, and building construction. These loan accounts appear on your balance sheet as liabilities. When you pay off a loan or sell a property, the loan balances for the property will return to 0. We'll explain this rather complicated process, step-by-step, later in this chapter in the section on Recording the Sale of a Property. In our example Chart of Accounts, we use 2400 Land Acquisition Loan, 2405 Land Development Loan, and 2410 Construction Loan.

Setting Up a Development Job

To track the costs and income for each individual building in a development, set up each one as a customer. For example, our Twice Right Construction Company is developing property on Ellis Street. Since we don't know who the customer is yet, we set up a customer called Ellis Street. If you have more than one job in a single development, you can make the development the customer and set up the properties as individual jobs. For example, if Twice Right purchased three lots on Ellis Street, the customer would be Ellis Street Development and each job would be its street address — 1120 Ellis Street, 1130 Ellis Street, and 1140 Ellis Street, as shown in Figure 19-1. However, in most cases you'll be using a lot and block number, i.e., Lot 1, Block 10 or L1B10.

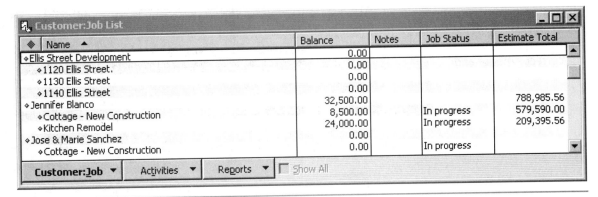

Figure 19-1
Notice, at the top of the Customer:Job List that Ellis Street Development is the customer and the jobs are listed as the address of each piece of property. If you don't have an address for the property, you may have to list each job as Lot 1, Lot 2, etc.

Using Items to Track Construction Costs as WIP

The methods we show you here are just for a spec builder or developer who is setting up and linking items (job phases of construction). This is different from the way we showed you to link items in Chapter 4. In this section, we'll use WIP and Construction Loan accounts.

If you want to track the costs of development by phase you need to create items and subitems for each phase. Now let's see how to set up a new item for 34 Land Dev (short for Land Development Costs), and a subitem 34.1 Loan Fee (short for Land Development Loan Fees) as shown in Figures 19-2 and 19-3. To create the new item:

▌ From the **Lists** menu, choose **Item List**. In the **Item List** window, pull down the **Item** menu and choose **New**.

▌ In **Type**, select **Service**.

▌ In **Account**, enter a Land Purchase, Land Interest/Closing Costs, WIP — Land Development, or WIP — Construction account. In our example, we selected 1480 WIP — Land Development because all the costs related to this item should get classified as land development costs.

Chapter 19: Real Estate Development

Figure 19-2
When you set up an item, be sure to link it to the correct expense and income accounts.

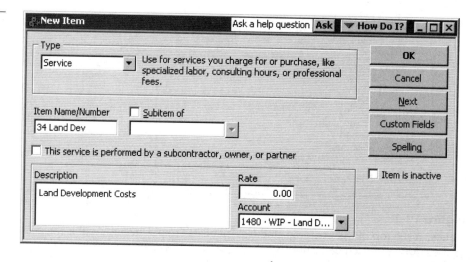

Figure 19-3
Here's an example of a new item that's a subitem. Notice that the "Subitem of" box is checked.

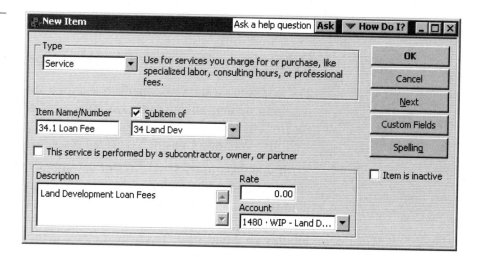

■ Fill in the other necessary fields and click **OK**.

Keep in mind that each item you set up as a spec builder needs to be set up with an asset account (in the Account box). Link land purchase items to a land purchase asset account such as 1470 Land Purchase. Link land development items to a WIP development asset account such as 1480 WIP Development. Link construction cost items to a WIP construction asset account such as 1490 WIP Construction and a construction liability loan account such as 2410 Construction Loan.

Use the same procedure to create subitems, remembering that you have to set up an item before you can set up any subitems under it. Figure 19-3 shows how we created 34.1 Loan Fee as a subitem of the 34 Land Dev item. Later in this chapter, we'll add a few more items that we'll use when the property is sold.

Land Purchase Transactions

Earnest Money

When you write a check to an escrow company to open escrow to purchase a piece of property, the money is called earnest money. In effect, it's a good faith deposit. Here's how to record the transaction and write an earnest money check:

- From the **Banking** menu, choose **Write Checks**. See our example in Figure 19-4.

- At the top of the Write Checks window, in **Bank Account**, make sure you select the checking account you write the check from.

- In **Pay to the Order of**, enter the name of the escrow company. Enter the check amount beside the company name.

- Make sure the check number is accurate or leave it blank if you're going to print the check later.

- In **Memo**, enter a description, such as land address, and a brief statement saying the check is for escrow earnest money.

- Click the **Expenses** tab.

- Click in the **Account** column and select your Escrow Deposit account from the account list. In our example, we used 1460 Escrow Deposit.

- In **Amount**, enter the full amount of the check.

- Leave Customer:Job and Class blank.

- Click **Save & Close**.

Chapter 19: Real Estate Development

Figure 19-4
Here's an example of how to record earnest money you pay to a title company to start escrow.

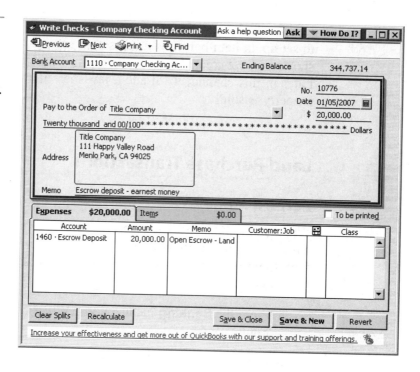

Closing Escrow on a Land Purchase

You'll need to write another check to the escrow company when you complete the land purchase and the title to the land is switched into your company's name. When you write this check, you can also record that you purchased the land and paid closing costs. And if you have a seller carry back loan or a loan from a bank, you can also record that on the check to the escrow company. Let's work through our example so you can see how to do this. In Figure 19-5, we show you an escrow closing sheet you can use to follow our example. Here are the steps:

- From the **Banking** menu, choose **Write Checks**. See our example in Figure 19-6.
- In the Write Checks window, in **Bank Account**, select the account that you're issuing the funds from.
- In **Pay to the Order of**, enter the name of the escrow company.
- In the **$** field, enter the amount of the check you write to the escrow company when the escrow on the land closes.
- Click the **Items** tab.
- Click the **Item** column and select your land purchase item from the pull-down list.

Figure 19-5

Here's an example of an escrow closing statement for a land purchase.

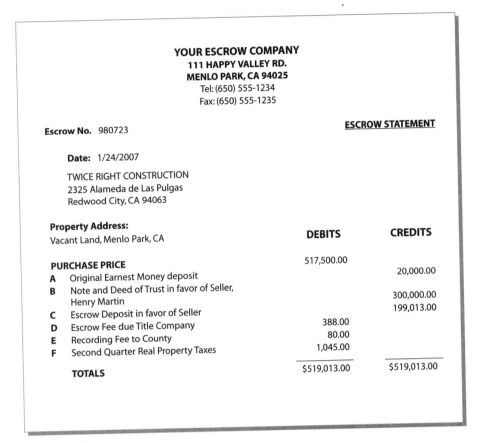

Figure 19-6

Use the Items tab to record land purchase and closing costs. In this example, we used items 31.1 Land Purchase Price and 31.2 Land Closing Costs. Your items for land purchase and closing costs may be different.

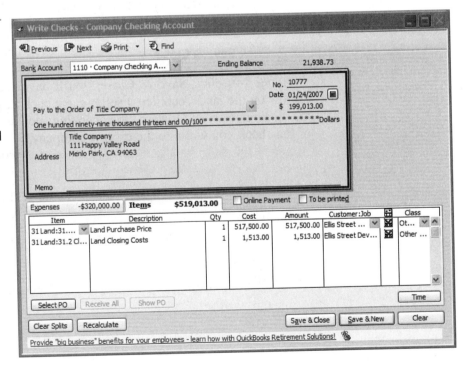

- In **Amount**, enter the purchase price from the escrow settlement sheet. In Figure 19-5, it's $517,500.

- Click the next line in the **Item** column and select your land closing costs item from the pull-down list.

- In **Amount**, enter the sum of all the closing fees (lines D - F in Figure 19-5).

If the seller carries back a loan, you record the loan amount on the Expenses tab. Here's how you can do that using the Expenses tab on the check shown in Figure 19-6:

- Click the **Expenses** tab. See Figure 19-7.

- Click in the **Account** column and select your Escrow Deposit account from the pull-down list. It's 1460 in our example.

- In **Amount**, enter the escrow earnest money as a negative number. In our example, it's −$20,000 (line A in Figure 19-5).

- Click the next line in the **Account** column and select the Land Acquisition Loan account from the pull-down list. It's 2400 in our example.

- In **Amount**, enter the total amount of the loan or seller carry back amount, as a negative number. In our example, it's −$300,000 (line B in Figure 19-5).

- Click the **Customer:Job** column and select the customer for the transaction from the pull-down list.

- Leave **Class** blank.

- Click **Save & Close**.

After you click Save & Close, you'll have recorded that you closed escrow on the land, recorded a seller carry back loan (if needed), recorded the closing costs, and lowered the checking account balance by the amount of the check you wrote to the escrow company.

Figure 19-7

Here's how to enter a seller carry back loan on land and how to clear out the earnest money involved in the sale.

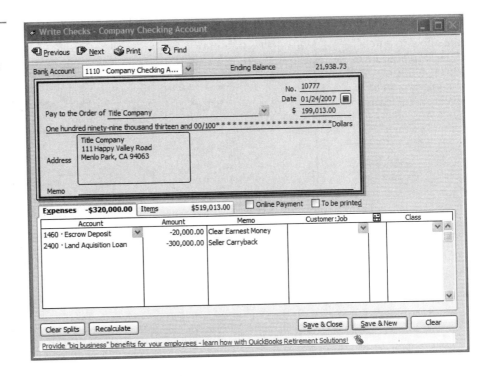

Personal Loans

Most real estate developers or spec home builders need to borrow money to buy land or do some developing before they take out a construction or development loan. These are basically personal loans. Whenever you get a personal loan (not including a construction loan) from a new entity, you need to add a new liability account to your Chart of Accounts. For example, let's say that our Twice Right Construction Company borrowed $25,000 from an investor. The investor is Paul Olsen, a long-time friend. We need to keep track of this money after we deposit it in our bank account. So we need to set up a new liability account for the loan.

However, if you don't have a loans payable header account in your Chart of Accounts, you should set that up first. In our example, we use 2300 Loans Payable as the loans payable header account. Then, we use the subaccount 2310 for Paul Olsen's personal loan to the business.

Chapter 19: Real Estate Development

Figure 19-8
Here's how to enter a new loan account. Don't use this method for construction or development loans.

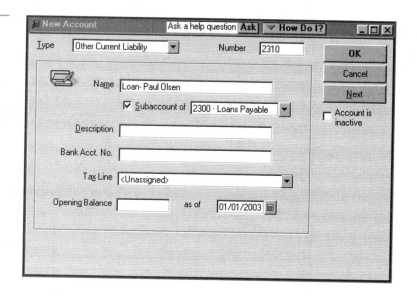

Now, on to the subaccount for the entity you borrowed money from. Keep in mind that this loan could be from an individual, another company, or a bank. In our example, the loan is from an individual, but the procedure would be the same if it were from another entity. However, don't set up a new account if the loan is a construction loan. Construction loans are handled differently and will be covered later in this chapter, under the heading Construction Loans.

- From the **Lists** menu, choose **Chart of Accounts**.
- In the Chart of Accounts window, pull down the **Account** menu and choose **New**.
- From the **Type** pull-down list, select **Other Current Liability**.
- In **Number**, enter the new account number. In our example it's 2310.
- In **Name**, enter Loan – name of entity (Paul Olsen in our example). See Figure 19-8.
- Click the **Subaccount of** box and select the loans payable header account number from the pull-down list. It's 2300 in our example.
- Click **OK**.

Follow the same procedure for any other entities that loaned you money. Keep in mind that you only enter new accounts for loans that are not construction or development loans. Construction, development, and seller carry back loans on land purchase are usually handled differently from personal loans. Construction loans, for example, have draw schedules and have to be paid back during the close of escrow. The personal loans we set up here are typically more informal than construction loans.

Figure 19-9
Here's how to record the deposit of a personal loan.

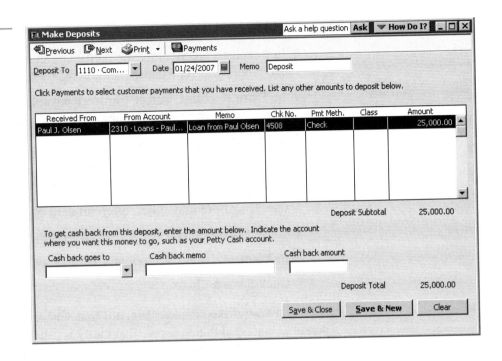

Depositing a Personal Loan

To record that you've deposited the money you received from a personal loan, you'd follow the instructions below. In our example, we deposit the $25,000 that Paul Olsen loaned us.

First, make sure that this deposit is separate from your ordinary income deposits. For example, if you received three checks that need to be deposited, two checks from customers and one personal loan check, make sure you use a separate deposit slip for the personal loan check. The other two customer checks can be added together on the same deposit slip, but the personal loan has to be deposited separately. This is because it will be recorded individually in the QuickBooks Pro checkbook reconciliation, and it would be best if the bank recorded it the same way.

- From the **Banking** menu, choose **Make Deposits**.
- At the Payments to Deposit window, select the payments you want to deposit.
- Click **OK**.
- In the Make Deposits window, from the pull-down **Deposit To** list, make sure you choose the checking account that you're going to deposit the money to. In our example shown in Figure 19-9, we use 1110 Company Checking.

- In **Date**, enter the date you'll be taking the deposit to the bank.
- In **Received From**, enter the entity you got the money from (not the job where you'll be using the money).
- In **From Account**, select the loan payable account from the pull-down list for the entity. In our example, it's 2310 Loans - Paul Olsen.
- In **Memo**, enter what you want to see in your checkbook register.
- In **Chk No.**, enter the check number from the entity.
- In **Pmt Meth.**, select from the pull-down list whether you received a check, cash, money order, or credit card payment.
- Leave **Class** blank.
- In **Amount**, enter the amount you're depositing.
- Click **Save & Close**.

Following these steps will record that you received and deposited the personal loan, and that you owe that entity for the personal loan. Whenever you print a Balance Sheet, the liability for this personal loan will appear, reminding you that you have to pay this personal loan back at some point.

Recording an Interest Payment on a Personal Loan

You may be required to pay interest on a personal loan on a periodic basis such as monthly, quarterly, or yearly. Or you may not have to pay both principal and interest until you pay back the personal loan. Regardless of when you pay the interest, the procedure is the same. You enter one transaction for the interest payment and one for the principal payment. However, the check you give to the entity may include just interest, or both interest and principal.

To record that you owe an interest payment on a personal loan (even if you may not be writing the check for some time) you'd enter a bill following the steps below. The bill you enter will track the interest expense, a job expense on the job reports (if a job is associated with the expense), and a transaction in your Accounts Payable (for interest only) to the entity you borrowed the money from.

- From the **Vendors** menu, choose **Enter Bills**.
- In the Enter Bills window, from the pull-down Vendor list, select the entity you owe the interest payment to. In our example shown in Figure 19-10, it's Paul Olsen.
- In **Date**, enter the date you owe the interest.
- If the entity has given you a bill or invoice for the interest, enter that number in the **Ref. No** field. In our example, we didn't have a bill from Paul Olsen, so we entered Interest.

Chapter 19: Real Estate Development

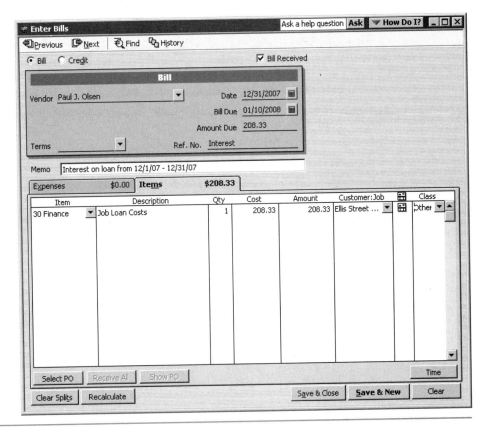

Figure 19-10
Here's how to record interest due on a loan.

- In **Amount Due**, enter the amount of the interest only.
- In **Memo**, enter a description for the bill.
- Click the **Items** tab.
- Click in the **Item** column and from the pull-down **Item** list, select the appropriate item.
- Enter **Amount, Customer:Job,** and **Class**. This will capture the interest cost on job cost reports.
- Click **Save & Close**.

Entering this bill will record that you owe the entity for the interest. The bill will show up in the Pay Bills window when it's due.

Contractor's Guide to QuickBooks Pro   **273**

Chapter 19: Real Estate Development

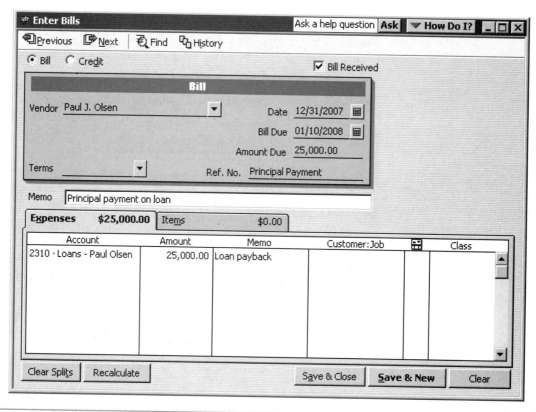

Figure 19-11
Here's how to record paying back a personal loan.

Recording a Principal Payment on a Personal Loan

When it comes time to pay down (or pay off) the personal loan, you'll need to follow these steps:

- From the **Vendors** menu, choose **Enter Bills**.

- In the Enter Bills window, from the pull-down **Vendor** list, select the entity you'll be paying the principal payment to. In our example shown in Figure 19-11, it's Paul Olsen.

- In **Date**, enter the date you decide to pay the principal.

- In **Ref. No**, enter the loan number or a description of the personal loan. In our example we didn't have a loan number, so we entered Principal Payment.

- In **Amount Due**, enter the amount of the principal only.

- In **Memo**, enter a description for the bill.

- Click the **Expenses** tab.

- Click in the **Account** column and select the account you set up when you borrowed the money. In our example, the account is 2310 Loan - Paul Olsen.
- Click **Save & Close**.

Entering this bill will record that you owe that entity for a principal payment. The bill will show up in the Pay Bills window when it's due.

Development Loans

A development loan is a loan you take out specifically to develop the infrastructure of a land parcel, such as subdividing the land into lots, sidewalks, roads, common areas, bridges, parks, etc. Usually the lender of a development loan requires you to also pay off the land loan as part of the development loan. So you also have to include the balance of the land loan in the amount you borrow for a development loan.

Closing Escrow on a Development Loan

There are several transactions that you need to record when you close escrow on a development loan. They are:

1. Record loan fees, including loan origination, document prep, processing, and underwriting fees.
2. Record escrow fees.
3. Record title insurance fees.
4. Record miscellaneous fees.
5. Record property tax payment.
6. Record paying off the seller carry back loan on the land. Skip this step if there's no loan on the land.
7. Record interest on the land loan. Skip this step if you don't have an existing land loan.

We'll take you through an example to show you how to do all this. You can use the Escrow Statement shown in Figure 19-12 to follow our example. To begin the process, we'll write a check like the one shown in Figure 19-13.

- From the **Banking** menu, choose **Write Checks**.
- In the Write Checks window, from the pull-down **Bank Account** list, select your Adjustment Register account if the development loan paid off the land loan. If you didn't pay off the land loan with the development loan and you wrote a check at the close of escrow, select the account you wrote that check from.

Chapter 19: Real Estate Development

YOUR ESCROW COMPANY
111 HAPPY VALLEY RD.
MENLO PARK, CA 94025
Tel: (650) 555-1234
Fax: (650) 555-1235

Escrow No. 123456-ABC   <u>**ESCROW STATEMENT**</u>

Date: 3/15/2007

TWICE RIGHT CONSTRUCTION
2325 Alameda de Las Pulgas
Redwood City, CA 94063

Property Address:
Vacant Land, Menlo Park, CA

		DEBITS	CREDITS
	YOUR ESCROW CO. Loan Proceeds		756,000.00
A	Loan Origination Fee	22,680.00	
B	Document Preparation Fee	650.00	
C	Processing Fee at 0.5%	2,340.00	
D	Fed Ex and Corp. Verif.	45.00	
E	Processing & Underwriting Fee	800.00	
F	Endorse (104.1) & Coll Set up	796.00	
G	Escrow Fee to Your Escrow Co.	250.00	
H	Title Insurance to Your Escrow Co.	1,266.50	
I	Inspection Fee	100.00	
	Record Documents	20.00	
	Reconveyance Fee	110.00	
	Recording Specific Guaranty	10.00	
J	Property Taxes paid to the County	4,914.97	
K	Seller carry back loan payoff		
	Principal	248,863.06	
L	Interest due at lot loan payoff date	5,178.20	
M	Funds Available	467,976.27	
	TOTALS	**$756,000.00**	**$756,000.00**

Figure 19-12
Here's an escrow statement you can use to follow our example.

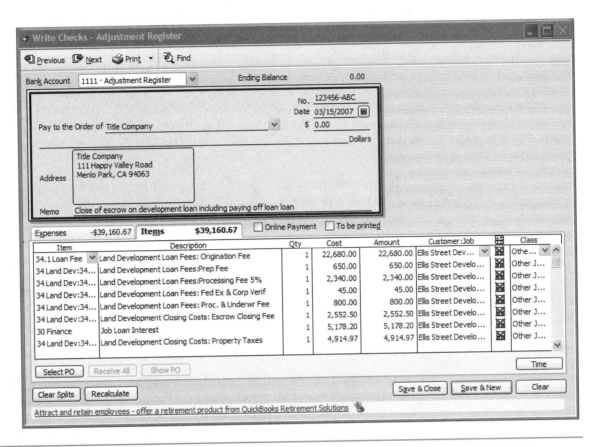

Figure 19-13
Use the Write Checks window to enter information from an escrow closing statement.

- In **$**, enter the amount of the check you wrote at the close of escrow. If you didn't have to write a check at the close of escrow because the new loan paid all of your costs, enter 0 as shown in the example. See Figure 19-13.

- In **No.**, enter the escrow number from the escrow statement.

- In **Date**, enter the date of the escrow statement.

- In **Pay to the Order of**, enter the name of your escrow company.

- In **Memo**, enter a description such as close of escrow.

Recording Fees and Taxes on a Development Loan

Now let's move on to enter the fees and taxes on the escrow closing. We'll go to the Items tab and enter lines in the Item column for these amounts. All the amounts entered here (except the land loan interest on line L) are charged to a subitem of 34 Land Dev Costs item. The amount on line L is the interest on the loan up to the escrow funding date. You charge it to item 30 Finance. If you have a loan on the land, you'll be required to pay this interest. In our example, we paid the seller $5,178.20 for interest through escrow.

In our example, the property taxes hadn't been paid for the prior year, so there was a penalty tax payment due (line J). Also, the taxes needed to be prepaid for the current year.

- Click **Items** tab.
- Click in the **Item** column and enter necessary amounts in the **Amount** column. In our example shown in Figure 19-13, we enter amounts from lines A through J shown in Figure 19-12.

Recording Paying off a Land Loan

Now we're ready to record the payoff of the loan on the land. We'll do that on the Expenses tab of the check shown in Figure 19-14. Since the full purchase price of the land was entered earlier, (see the section titled Closing Escrow on a Land Purchase which shows how we entered the full land value shown in Figures 19-5 and 19-6), all we want to do here is record that we paid off the land loan. In our example in Figure 19-12, the seller carry back land loan balance was $248,863.06.

- Click the **Expenses** tab. See Figure 19-14.
- Click in the **Account** column and enter your land acquisition loan account. In our example, it's 2400 Land Acquisition Loan.
- In **Amount**, enter the total amount of the loan payoff. In our example shown in Figure 19-14, the amount is $248,863.06.
- If the development loan paid these costs, you need to select account 2405 Land Development Loan and enter as a negative number the total of the fees (from the Expenses tab) and the amount the bank paid on your behalf to pay off the land loan. This will increase the loan balance.
- Click **Save & Close**.

The total amount of the check should be zero *or* if you wrote a check at the close of escrow to cover some costs, the check should be made out for the amount you paid at the close of escrow.

Chapter 19: Real Estate Development

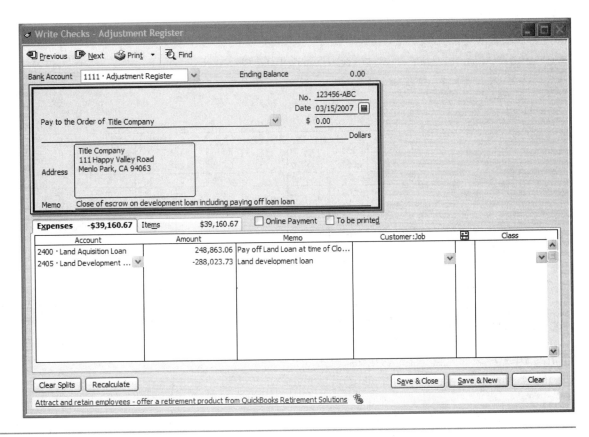

Figure 19-14
To enter an amount from an escrow closing statement use the Expenses tab of a Write Checks window.

Construction Loans

Entering a Construction Draw

When you draw funds for a development project, you need to create an invoice before you deposit the funds. And before you create the invoice, you need to create an estimate for the project. Look back to Chapter 12 if you need help with estimates. Here's how to enter an invoice for a construction loan draw:

▌ From the **Customers** menu, choose **Create Invoices**.

▌ In the Create Invoices window, select the **Customer:Job** you're requesting the funds for. In our example, we selected Ellis Street Development — 1120 Ellis Street.

Contractor's Guide to QuickBooks Pro **279**

Chapter 19: Real Estate Development

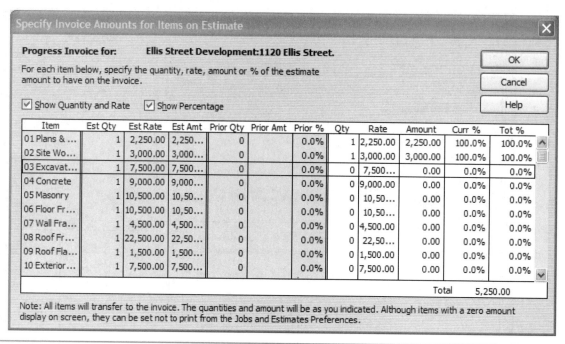

Figure 19-15
Use the Specify Invoice Amounts for Items on Estimate window to enter the amount or % for each line item of the draw.

- If you entered an estimate for this job, you'll see the Available Estimates dialog box. Select the estimate you would like to bill from.

- Click **OK**.

- In the Create Progress Invoice Based on Estimate window, select **Create an invoice for selected items or for different percentages of each item**.

- Click **OK**.

- In the Specify Invoice Amounts for Items on Estimate window, in **Amount**, enter the amount you're billing for. In our example shown in Figure 19-15 the draw is for $5,250 for item 01 Plans and item 02 Site Work.

- Click **OK**.

Back in the Create Invoices window, make sure the correct amounts appear in the draw invoice. Since each job phase item was originally linked to the Construction Loan account, when this invoice gets generated the item will automatically track the draw as a liability.

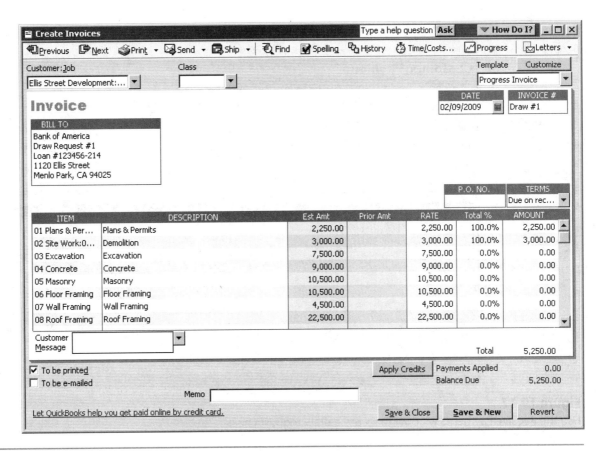

Figure 19-16
Here's how to record a percent complete construction draw.

> In **Bill To**, enter the bank or investor you're requesting the funds from. In our example, shown in Figure 19-16, we selected Bank of America.
>
> Leave **Class** blank.
>
> Click **Save & Close** to create the invoice.

Depositing the Draw Check into Your Account

Here's what you need to do when you receive the draw check from the bank:

> From the **Customers** menu, choose **Receive Payments**.
>
> From the pull-down **Received From** list, select the same customer that's on the draw request. You should see the original draw request appear, as shown in Figure 19-17.

Chapter 19: Real Estate Development

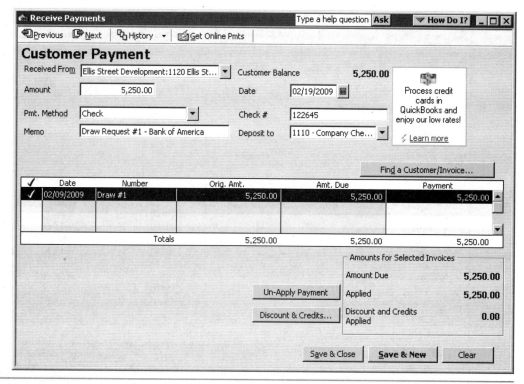

Figure 19-17
Use the Receive Payments window to record a draw request check.

- Fill in the date you received the payment, the amount paid, the payment method, and the check number if paid by check.
- In **Memo**, enter the draw request number and the bank or investor you requested funds from.
- Click **Save & Close**.

Recording the Sale of a Property

When you sell the property, you need to:

1. Record the sale of the property.
2. Record paying off the construction loan.
3. Account for the closing costs.
4. Transfer the WIP accounts to the Cost of Goods Sold account.

This looks complicated, but we'll show you how to do it all in just one transaction. The transaction will be on the invoice you issue for the sale of the property. Shortly, we'll walk you through the process of entering the invoice in QuickBooks Pro. First, you need to set up these seven items that you need for the process. Use the usual method to set up these items. Make sure you link the items to the proper accounts.

- *Building Sale* — link to your Construction Income account (4110 in our Chart of Accounts)
- *Building Costs* — link to your Job Related Costs account (5110 in our Chart of Accounts)
- *WIP Trans* — link to your WIP Construction account (1490 in our Chart of Accounts)
- *Closing Costs* — link to your Job Related Costs account (5110 in our Chart of Accounts)
- *Const. Draws* — link to your Construction Loan account (2410 in our Chart of Accounts)
- *Land Cost* — link to your Job Related Costs account (5110 in our Chart of Accounts)
- *Land Int* — link to your Job Related Costs account (5110 in our Chart of Accounts)

Recording an Invoice for the Sale of a Property

Now you're ready to create an invoice to record the sale and closing costs, and to clear the WIP assets and Construction Loan accounts. For our example, we created an invoice for 1120 Ellis Street when we received a check from the title company that closed the escrow on the new house. The invoice is shown in Figure 19-18. Let's take a close look at each item in the figure:

- *Building Sale* — This is the sale price of the property. Amount is the selling price of the home.
- *Land Cost* — Records the purchase price of the land in a cost of goods sold (job related costs) account.
- *31.1 Land item* — This line clears the purchase price of the land out of the asset account. If you purchased several parcels in a subdivision, then the price is a portion of the original purchase price. For example, in Figures 19-5 and 19-6, we purchased land that was subdivided into five lots. Each lot was similar in size, so we divided the $517,500 by 5 to get an individual lot price of $103,500.
- *Land Int* — Records the land interest and closing costs in a cost of goods sold (job related costs) account.
- *31.2 Land Int item* — This line clears out the land closing costs and interest through the end of the life of the land loan and divides the costs by the five lots. In our case, we paid $1,513 for closing costs plus $23,450 in interest. Divide that by the five lots and you have $4,992.00 per lot.

Chapter 19: Real Estate Development

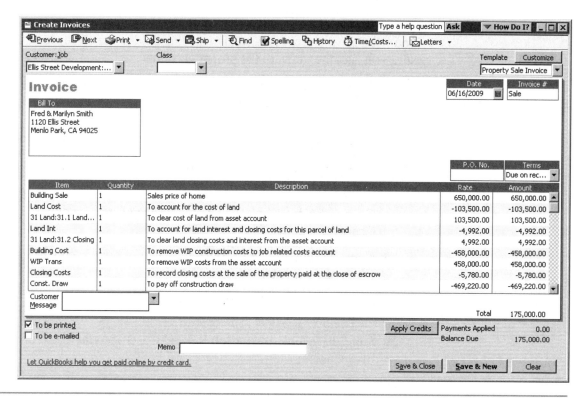

Figure 19-18
Use the Create Invoices window to record the sale of a property.

- *Building Cost* — This moves the WIP costs to the Cost of Goods Sold account. Amount is the same as the amount in the next line except it's a negative number.

- *WIP Trans* — This line clears the WIP Construction account total out of the asset account. Amount is the total WIP costs for the project. To figure out the total WIP costs for the job, print out a Profitability Detail Report. Chapter 17, Reports, has more information on this.

- *Closing Costs* — Amount is the total of the real estate commission fees, recording fees, title fees, and any other credits given to the buyer at the close of escrow for the sale of the property. This is also a negative number because the seller won't receive this money; it's been taken out of the funds to be disbursed to the seller (contractor).

■ *Const Draw* — This is the amount the escrow company paid to the lender for the total amount of construction draws taken to date. This total will be listed on the closing papers as the loan payback. It's a negative number because it's being taken out of funds to be disbursed to the seller (contractor). This line pays back the construction loan.

The total amount of the cash sale needs to be exactly equal to the amount of money received at the close of escrow.

Recording a Title Company Payment for the Sale of a Property

To record the payment from the title company for the sale of a property:

■ From the **Customers** menu, choose **Receive Payments**.

■ From the pull-down **Received From** list, select the same customer that's on the invoice for the sale of the property.

■ In the list, select the invoice you received payment for.

■ Fill in the date you received the payment, the amount paid, the payment method, and the check number if paid by check.

■ Click **Save & Close**.

To record the payment after depositing it in the bank:

■ From the **Banking** menu, choose **Make Deposits**.

In this chapter, you've learned how to set up QuickBooks Pro to track each transaction using proper accounting methods for a spec builder. This chapter wasn't meant to be read just one time. In fact, this book wasn't meant to be read only one time. We hope you'll use this chapter and the whole book as a reference guide. For example, you can use this chapter to help set up and link items to track WIP and construction loans. When you sell a property, you'll need to refer back to this section and follow the instructions on how to record the sale of the property. When you receive the money, you can go through receiving the payment. Then when you deposit the money to the bank, you can follow our instructions again.

Conclusion

It's rare that we run across a contractor or builder who enjoys accounting, let alone job costing, payroll, workers' comp reports, or doing income taxes. If you enjoyed these things, you'd be an accountant, bookkeeper, or consultant (as we are). But if you've followed the instructions in this book, you're on your way to having an efficient and accurate accounting and job costing system. And the bottom line is the bottom line — with accurate reporting, you'll be more profitable.

By now you realize that setting up QuickBooks Pro isn't an easy task. We've tried, however, to break up the tasks into manageable units that are simple to understand and duplicate. If you followed our directions in setting up QuickBooks Pro, and were diligent in your efforts to enter transactions accurately, you'll be rewarded with clear and useful job cost reports — the Holy Grail of construction accounting. Of course, there are other important accounting reports you can get from QuickBooks Pro. And we encourage you to continue to explore them as they apply to your business. Craftsman Book Company (www.craftsman-book.com) publishes several "accounting for contractors" books, including *Markup & Profit* and *Builder's Guide to Accounting*, which will help you apply wisdom to your numbers. There are also several on-line sources you can use to increase your construction accounting knowledge. For nonspecific accounting problems, Intuit, the creator of QuickBooks Pro, has its own Web site (www.intuit.com). For QuickBooks Pro help specific to the construction business, take a look at www.onlineaccounting.com.

As you work with QuickBooks Pro, you'll be amazed at its flexibility and power. We don't suggest that you turn a blind eye to new solutions. No one has all the answers, including us. We're constantly finding new solutions and "work-arounds" to make the process smoother. Also, everyone puts different emphasis on the value of individual reports, so one solution may not sit right with everyone. Remember, if the information is what you need, use it. Don't be misled by "pundits" that are selling the "right way."

In writing this book, we've tried to create a system or process that you can follow on a daily basis. We urge you to view accounting and job costing as a repeatable procedure that will get simpler as you do it over and over again. Repeatability works for you in several ways. It cuts down on mistakes and helps you refine procedures for maximum productivity. It also makes it easier to train a new employee.

We also want to touch a little on the future of computerized construction accounting. There's no doubt in our minds that in the future, the Internet will play a major role in your business. You'll be estimating, purchasing, and paying bills on-line — it's not a question of if, but when. And Intuit is at the forefront of on-line accounting initiatives. This type of "e-commerce" will only get easier to do, cheaper to implement, and more popular. The paperless transaction is already here. For example, you no longer get a receipt when you charge something on your credit card over the phone. The construction trades are sometimes slow to catch on, but nevertheless, it won't be long before invoices from your subs and suppliers will go directly into QuickBooks Pro without you ever receiving a paper invoice. And they'll even be earmarked with job costing codes so one day you'll turn on your computer and simply ask for a job cost. Without any input — the data will simply be there. Until then, we'll sit at our computer terminals, wishing we didn't have to code this invoice or break down that timecard.

Appendix A

Estimating with QuickBooks Pro

In Chapter 12 we suggested that any contractor who has a good estimating program and likes it should stick with that program rather than switch to QuickBooks Pro estimating. You can create estimates outside of QuickBooks Pro and enter only section summaries (for example, the job phases in *sample.qbw*) into QuickBooks Pro. We know that works because we know many contractors who do exactly that. But you need to know that there's another way. In this appendix, we'll explain how to use the estimating functions built into QuickBooks Pro.

There's a lot to like about QuickBooks Pro estimating. It works. It comes free with QuickBooks Pro. And, like all QuickBooks, it's very forgiving. Practice all you want. Don't worry about making mistakes. It's easy to undo almost anything.

Three Good Reasons to Try QuickBooks Pro Estimating

First, QuickBooks Pro estimating is the starting point for all the accounting and job cost tracking features built into QuickBooks Pro. Once you've created an estimate, click to turn that estimate into an invoice. The invoice amount becomes an entry in your accounts receivable and an asset on your balance sheet. You can compare the actual cost of doing each part of the job with your estimates. The difference between costs and receipts shows up as your profit. That's a good, professional way to do business. Every construction company should be run that way. And it all starts with a QuickBooks Pro estimate.

Second, the progress billing features in QuickBooks Pro can be a huge advantage if you send out detailed bills for the work done to date and have to keep track of what's been paid and what's still collectible. The only way to do progress billing in QuickBooks Pro is to start with a QuickBooks Pro estimate.

Progress billing straddles two topics, estimating and billing. We're not going to say much more about progress billing in this appendix. Actually, there isn't much to say. The estimating part of progress billing is easy to learn. You'll find more information on this in Chapter 13, Receivables.

Third, your invoice, like your bid or estimate, should identify each cost in the job. The best invoice for construction work is a perfect image of the bid, repeating each item in the bid and showing the same costs. Why create an estimate and then retype a separate invoice? That's duplicate work. QuickBooks Pro converts estimates to invoices at the click of your mouse button.

Most of what we're going to explain in this chapter isn't described in the QuickBooks Pro User's Guide. But if estimating is important to you, keep reading. We're going to cover it all.

A Road Map to Your Destination

You probably know already that QuickBooks Pro doesn't come with any construction cost estimating data. And for good reason. QuickBooks is written to work as well for a dry cleaning business, or a barber shop or a car wash, as for a construction contractor. No wonder you won't find the cost of hanging drywall anywhere in QuickBooks Pro. You have to create every labor and material cost estimate in your database by typing it into the Item List.

Building that Item List manually takes time. It's no problem to type a few dozen descriptions and costs into the QuickBooks Pro Item List. But an Item List suitable for detailed labor and material cost estimates might require several thousand lines. Even after building a long Item List, you may not be satisfied. Finding a particular item on a list of thousands can take a lot of scrolling up and down. There's no Item List index in QuickBooks Pro. You can sort items alphabetically. But items are only alphabetical within each item type, such as Non-inventory Part, Service, or Other Charge.

And there's one important limit built into the QuickBooks Pro Item List. An item name can't be longer than 31 characters. This truncating can be a handicap when you're describing something like "⅝-inch type X gypsum wallboard screwed in place, taped, textured and painted two sides." That's 88 characters, much too long for a QuickBooks Pro item name. So you have to abbreviate, such as GWB5/8XSTTP2S. Not many estimators are eager to deal with several thousand codes like that.

In Appendix B we'll show another way, creating an estimate outside of QuickBooks Pro (using the National Estimator program and ready-made database on the CD in the back of this book) and importing the completed estimate into QuickBooks Pro. The advantages should be obvious. But please don't skip Appendix A just because you prefer doing what's described in Appendix B. Every contractor using QuickBooks Pro should understand QuickBooks Pro estimating. Walk before you run. When you're comfortable with the information in this appendix, move on to Appendix B.

Setting Preferences for Estimating

To begin, let's be sure we're reading off the same sheet by checking your preference settings. See Figure A-1.

- From the **Edit** menu, choose **Preferences**.
- Click the **Jobs & Estimates** icon and the **Company Preferences** tab.
- Answer **Yes** to the question **Do You Create Estimates?**
- For now, answer **No** to **Do You Do Progress Invoicing?**
- Scroll down the Preferences list on the left side of the window. Click the **Sales Tax** icon and then the **Company Preferences** tab.
- Be sure to answer **Yes** to **Do You Charge Sales Tax?** See Figure A-2.

If you collect sales tax and haven't set up the most common sales tax and its rate, you'll need to do that now.

- From the pull-down Most common sales tax list, select **Add New**.
- In the New Item window, enter the type of tax (probably Sales Tax Item), name of the tax, description, tax rate, and taxing agency.
- If you haven't set up the taxing agency, you'll need to do that now. From the pull-down Tax Agency (Vendor) list, select **Add New**.
- In the New Vendor window, enter information about the taxing agency.
- Click **OK** in the New Vendor window to return to the New Item window.
- In the New Item window, if you've entered all necessary information, click **OK**.

Back in the Sales Tax Preferences window:

- Don't select Mark taxable amounts with "T" when printing unless you want the taxable amounts on estimates marked with a T in the Tax column.
- Click **OK**.

Appendix A: Estimating with QuickBooks Pro

Figure A-1
Use the Company Preferences tab of the Jobs & Estimates Preferences window to tell QuickBooks Pro you want to create estimates.

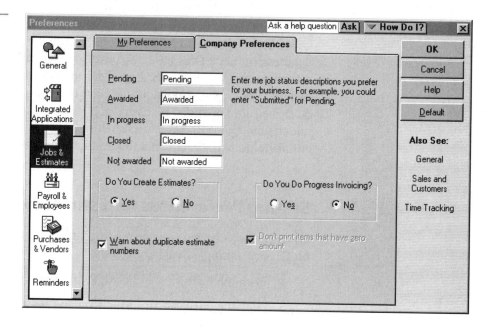

Figure A-2
Enter sales tax information in the Company Preferences tab of the Sales Tax Preferences window.

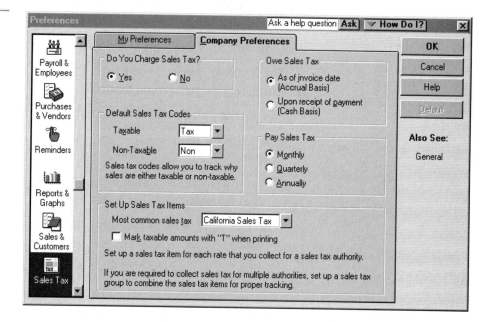

Building Your Item List

Figure A-3 shows a small estimate for flooring work that's part of an attic conversion job. Let's take a closer look at this estimate. Then we'll actually create the estimate. In the Item column on the estimate, you'll see three items — Vinyl Tile, Install Vinyl Tile, and Tile Flooring. Since QuickBooks Pro doesn't come with any cost estimating data, we'll begin by adding these items to our Item List.

We recommend that you distinguish between taxable and non-taxable items in your Item List. In many states, the materials you install are taxable to your customer. There's usually no tax on labor. If you collect tax from your customers and have to remit that tax to a state agency, your estimates and invoices have to identify what's taxable and what's not. Some states and Canadian provinces require that you keep a record of the tax collected on each invoice and write a check to the taxing authority when tax payments are due. If your state places a tax on material alone, you have to keep material costs separate from labor costs on your Item List. Keep that in mind as we create new items in the next section.

Even if you don't have to charge sales tax on material, it's good practice to separate labor and material costs on your Item List. That lets you compare estimated material costs and estimated labor costs with actual material costs and actual labor costs. Of course, for subcontracted work, you have only one lump sum — including both material and labor. There's usually no reason to separate labor and material costs for subcontracted work.

Figure A-3
Use the Create Estimates window to enter a description, quantity, and cost for each item you're including in an estimate.

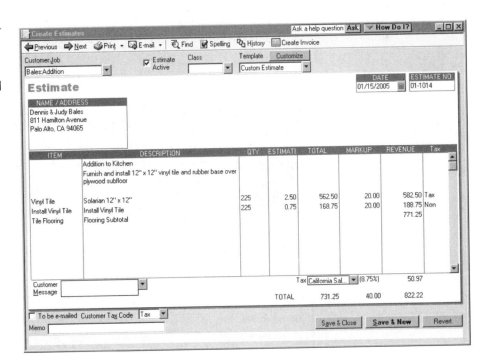

Appendix A: Estimating with QuickBooks Pro

Creating a New Item to Use on Estimates

There are two ways to build an Item List and we'll cover both. You can create items before you use them on an estimate. We'll demonstrate this first. Or you can create items as part of the estimating process. We'll demonstrate that when we begin writing the estimate.

To add a new item to the Item List:

- From **Lists** menu, choose **Item List**.
- In the Item List window, click **Item** and select **New**.
- From the pull-down Type list, select **Non-inventory Part**. QuickBooks Pro doesn't have a "material" category on the Item List. "Non-inventory Part" is as close as we're going to get. For subcontract costs, or if you prefer combining labor and material costs into one lump sum, select **Other Charge** from the pull-down Type list. We recommend using the Service category for labor costs exclusively.
- In **Item Name/Number**, enter the name of the new item. We entered Vinyl Tile as shown in Figure A-4.
- Select **This item is purchased for and sold to a specific customer:job**.
- In **Description on Purchase Transactions**, enter an appropriate description. We entered Solarian 12" x 12". What you enter will also appear under Description on Sales Transactions.
- In **Cost**, enter the cost of the item. Our cost per tile is $2.50.
- From the pull-down Expense Account list, select any Cost of Goods Sold account that makes sense to you. We used 5110 - Job Related Costs.

Figure A-4
Use the New Item window to enter a non-inventory part type item.

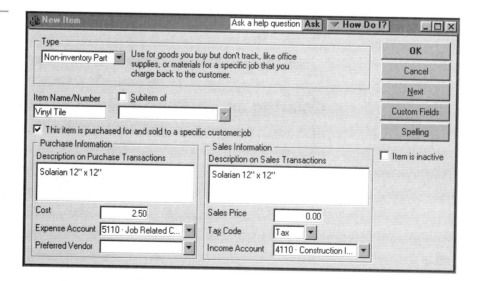

292 Contractor's Guide to QuickBooks Pro

- Leave Sales Price at 0.00.
- Select **Tax Code** if necessary. In our example, the vinyl is taxable.
- From the pull-down Income Account list, select any income account that makes sense to you. We used 4110 - Construction Income.
- Click **OK**.

A few words about what we've just done.

First, we've entered a cost price but not a sales price. But you don't have to enter a sales price. That's optional. You may prefer to leave the price open until you use the item in an estimate. Your selling price can change a lot. It's usually easier to enter the correct price when you use the item in an estimate than it is to keep changing prices on the Item List.

Second, you can enter both a cost and a selling price for an item. The difference between cost and selling price is markup, and markup usually varies by job, not type of material. We recommend leaving the sales price at 0.00 when you create a new item on the Item List.

Third, we've selected Non-inventory Part as the item type for the Vinyl Tile. If you have an inventory of vinyl tile and want to keep track of how much is in stock, there's another choice. Select Inventory Part as the item type if you intend to record each addition to inventory and sale from inventory. QuickBooks Pro will keep track of your tile inventory. But first you have to set the Purchases & Vendors Preference to "Inventory and purchase orders are active." For more information on preferences, see Chapter 2.

Since every cost item in your estimate has to be on your Item List, you'll get a warning if you try to enter an item that's not on the list. But that's not as inconvenient as you might expect. QuickBooks Pro lets you set up the item without leaving the estimate. We'll demonstrate this shortly.

Create another item for "Install Vinyl Tile." This is labor and should be entered as a "Service" type item. We'll create a subtotal item for "Tile flooring" during the estimating process to demonstrate how you could set up a new item while writing an estimate.

Creating an Estimate in QuickBooks Pro

Let's assume that you've set up the labor and material costs you use most often. Now it's time to start an estimate:

- From the **Customers** menu, choose **Create Estimates**.
- From the pull-down Customer:Job list in the Create Estimates window, select the customer and job. If you need help in entering a new customer, see Chapter 7, Customers and Jobs.

- Click in the **Description** column and type a short job title, such as "Addition to Kitchen." Don't press Enter unless you intend to insert a blank line below the description you just typed. Then describe the work you plan to do. See Figure A-3.

- On the line below the description you just entered, click in the **Item** column, and from the pull-down Item list, scroll down until you find the item you need. In our example, we selected Vinyl Tile. Description and price will fill in automatically.

- In **Qty**, enter the quantity. Amount will fill in automatically. In our example, we entered 225.

- In **Markup**, enter the markup. In our example, it's 20%.

- Use the same procedure to add another item for Install Vinyl Tile. Use the same quantity and markup.

Adding a New Item While Creating an Estimate

Here's how to add a new item to the Item List as you create an estimate:

- Click in the **Item** column just below the line "Install Vinyl Tile."

- Click the down arrow and from the pull-down list, select **Add New**.

- In the New Item window, from the pull-down Type list, select **Subtotal**.

- Enter **Item Name/Number** and **Description**. In our example shown in Figure A-5, we entered Tile Flooring and Flooring Subtotal.

- Click **OK** to return to the Create Estimates window.

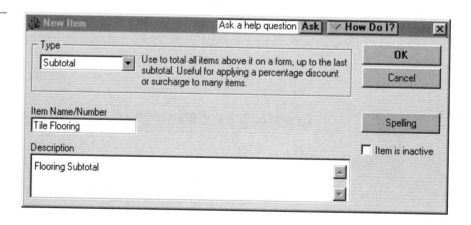

Figure A-5
Use the New Item window to enter a subtotal type item.

Grouping Labor and Material Costs

Once you have both labor and material items for some tasks on the Item List, consider grouping the two together so you can use them as one item on an estimate. Let's bundle the vinyl tile material and tile installation labor to create a group we'll call Solarian tile.

- In the Create Estimates window, from the pull-down Item List in the **Item** column, select **Add New**.

- In the New Item window, from the pull-down Type list, select **Group**.

- In **Group Name/Number**, enter the group name. In our example shown in Figure A-6, we entered Solarian 12" x 12" tile.

- In **Description**, enter the description of the group item. In our example, we entered Installed 12" x 12" Solarian tile.

- Click **Print items in group** if you want all the items in the group printed on an estimate.

- In the **Item** column, select each item to include in the group from the pull-down Item List. In our example, we selected Vinyl Tile and then Install Vinyl Tile.

- Click **OK** to complete the group and return to the Create Estimates window.

To use a group item:

- In a Create Estimates window, from the pull-down Item List, select the group item. Note that you still have to enter a quantity for each item in the group.

Figure A-6
Use the New Item window to select items to group together.

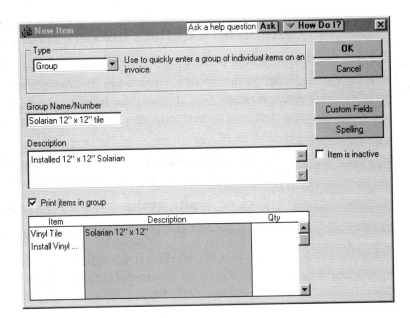

Item groups are most useful when many components are used together. For example, a group "Wall assembly" might include studs, plates, blocking, nails, ½-inch gypsum wallboard, paint, and the associated labor cost. Unfortunately, a group can't include another group, so you have to create your groups one at a time.

Adding a Tax Due Item in a QuickBooks Pro Estimate

Handling tax is another place where QuickBooks Pro really shines. By default, taxable items will have the word Tax in the Tax column at the right hand side of the estimate. In Figure A-3, you'll see Tax in the row Vinyl Tile. The tax total of $50.97 for the job is at the lower right of the estimate.

QuickBooks Pro will keep track of sales tax due in each state or county where you do business. If you need to, you can create an item on your Item List for each state or county where you file a sales tax report. To do this:

- From the pull-down Tax list near the bottom of the Create Estimates window, select **Add New**.
- Enter the tax name, description, rate, and tax agency.
- Click **OK** to return to the Create Estimates window.

The tax due will be calculated automatically for each taxing authority you set up. Of course, you have to select the correct taxing authority when creating estimates. When the time comes to pay the tax, choose the Pay Sales Tax command on the Vendors menu to see how much you owe each taxing authority.

Our sample estimate is complete now, so we'll:

- In the Create Estimates window, click **Save & Close**.

Turning an Estimate into an Invoice

QuickBooks Pro estimates don't have any effect on account balances. You can do all the estimates you want without changing anything on the company profit and loss statement or balance sheet. That's no longer true when you turn an estimate into an invoice. Estimates are just talk. Invoices are real money. To turn an estimate into an invoice:

- From the **Customers** menu, choose **Create Estimates** and select the estimate. We'll select our flooring estimate in Figure A-3.
- Click the **Create Invoice** icon at the upper center part of the window.
- Click **OK** and you'll see the invoice. Figure A-7 shows our invoice.

Notice that the Markup column is gone. In an invoice, the Amount column includes markup. So the balance due on an invoice will be the same as the estimate total. Only the way of arriving at that total has changed.

Figure A-7
An invoice created from an estimate.

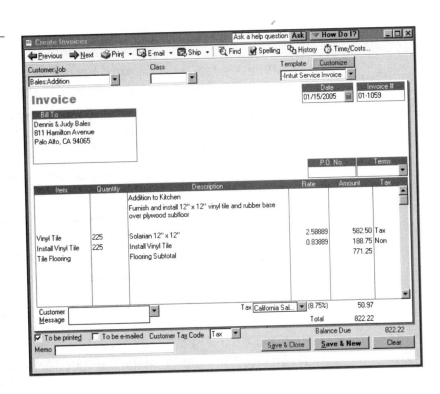

Let's take a tour around the invoice. You can change anything now. The invoice isn't official until you click on Save & Close.

- At the upper center of the Create Invoices window, consider changing Form Template to Intuit Service Invoice. That's more appropriate for a construction contractor.

- Check to be sure the Invoice #, Date, P.O. No., and Terms are what you want.

- In the memo box at the lower left of the estimate, type a short message if you want to.

- Check all the prices and descriptions to be sure they're correct. You can change invoice prices, quantities, or descriptions here. You can even add or delete lines at this point. Click the line on the estimate that you want to change or where you want to insert a line. Then click on **Edit** and Insert Line or Delete Line.

- When the invoice is perfect, print it. Then click **Save & Close**. The Create Invoices window will close after offering to check the spelling on the invoice.

When you click Save & Close in the Create Invoices window, you make major changes throughout your accounting system. Remember, an invoice represents money, and the primary job of QuickBooks Pro is to keep track of money. Let's look at some of the changes you made by clicking Save & Close in the Create Invoices window.

The Invoice in Your Accounting System

Totals from the new invoice will show up throughout QuickBooks Pro reports. You can check these reports to see for yourself.

- From the **Reports** menu, choose **Company & Financial**. The Profit and Loss reports show the income. The Balance Sheet reports show Accounts Receivable and the sales tax payable.
- From the **Reports** menu, choose **Customers & Receivables**. The A/R reports show the receivable.
- From the **Reports** menu, choose **Sales**. The Sales reports show the sale.
- From the **Reports** menu, choose **Jobs, Time & Mileage** and then **Jobs Estimates vs. Actuals Detail** to show the cost and revenue.

That one little invoice has worked its way throughout your QuickBooks Pro company data file.

Now, here's the part we like. QuickBooks Pro is so forgiving. Since we were just practicing with the Bales job, let's delete the invoice and see what happens.

- Close any reports you have open.
- From the **Customers** menu, choose **Create Invoices** and then **Previous** to get back to the Bales invoice.
- With the Bales invoice on your screen, click **Edit** and **Delete Invoice**.
- Click **OK** and the invoice is gone, not just off your screen, but throughout your QuickBooks Pro company data file. There's no trace of the Bales invoice anywhere in the QuickBooks Pro reports.
- From the **Customers** menu, choose **Create Estimates** and click **Previous** to search for the Bales estimate. You'll find it's still there. Deleting the invoice had no effect on the estimate. That means you could recreate the invoice once again from the estimate if necessary.

That's only part of the power in QuickBooks Pro. Any time before the invoice is paid, you can add lines, delete lines, or change anything you want. That's a nice feature. But, as a practical matter, once an invoice has been mailed to a client, it's better to issue a credit or create another invoice if you have a lot of changes to make.

Tidying Up Your Company File

When is it time to delete an invoice (or estimate)? Don't delete any estimate just because you've turned it into an invoice. If you delete an estimate too soon, you'll remove all estimated costs from the Job Estimates vs. Actuals reports. You need that information at least until you close the job and pay all the bills. You can usually delete any estimates that weren't accepted. That'll help reduce clutter in your company file.

Keep invoices (even after paid in full) at least until the end of the fiscal year during which there was any activity.

Here's the best rule to follow. Don't delete anything in your QuickBooks Pro company data file just because it's old and (you hope) no longer needed. Better to let QuickBooks Pro decide what isn't needed.

Here's how to clean up a company file:

- From the **Windows** menu, select **Close All**.
- From the **File** menu, select **Archive & Condense Data.**
- In the Archive & Condense Data dialog box, select **Condense transactions as of a specific date**. See our example in Figure A-8.
- Enter the date. It's good practice to save data in the current year and prior year.
- Follow instructions on the screen.

That's about all there is to know about estimating with QuickBooks Pro. In Appendix B, we'll go into more depth and show you how to prepare detailed estimates using National Estimator (a program on the CD in the back of this book) and then import those estimates into QuickBooks Pro.

Figure A-8
Use the Archive & Condense Data window to select categories you want QuickBooks Pro to consider removing.

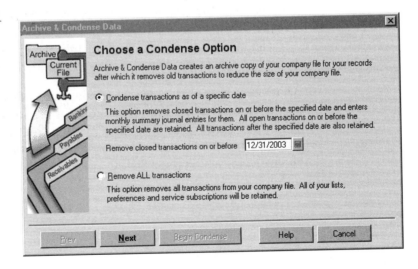

Appendix B

Job Cost Tracking and Importing Estimates

A Different Approach

This appendix is different from what you find in the first 19 chapters. Until now, we've covered the basics. And we've steered clear of topics some could find confusing or frustrating. This appendix is intended more for specialists, construction contractors who:

- Consider construction cost estimating an important task.
- Need a database with hundreds of pages of construction costs estimates.
- Write estimates that include dozens or hundreds of cost items.
- Want to compare estimated costs and actual costs when the job is done.
- Don't want to re-type their estimates into QuickBooks Pro.

Appendix A covered estimating as it exists in QuickBooks Pro. A major disadvantage is that QBP doesn't include cost estimating data. You have to build the database and keep that data current. There's another choice, and that's the subject of this Appendix B. The disk in the back of this book has several hundred pages of labor and material costs for construction, plus a program that makes it easy to compile estimates using this data. Once you've built an estimate, it exports directly to the estimating module in QuickBooks Pro, creating an Item List automatically. That's cool. It can save hours on any estimate.

Hundreds of construction cost estimating programs are available. A few export estimates to QuickBooks Pro. But none is used more widely than Craftsman's National Estimator, the program on the disk in the back of this manual. Over 60 estimating databases work with the National Estimator program. All are revised annually and cost about $50, including the entire database as a printed book. An order form is at the back of this manual. If you don't need a printed book, all Craftsman estimating databases are available by Web download for much less than $50. Point your browser at http://costbook.com and click on Try It/Buy It. These downloads are guaranteed to work perfectly on your computer. There's no charge for phone support. Call 760-438-7828 Monday thru Friday from 8 AM to 5 PM (Pacific).

Estimates created with any version of National Estimator export smoothly to QuickBooks Pro. That's what we're going to explain in this chapter — step-by-step.

Tracking Job Costs

When you use QuickBooks Pro to write a check for a construction expense (payroll, supplies, subcontracts), you can charge the expense to some cost category in the job estimate. See the five columns in the Estimates vs. Actuals Detail report for the Palmer Job on the next page.

- *Column 1 (Est. Cost)* — Estimated costs by job phase for material, labor, equipment, and subcontract expense.

- *Column 2 (Act. Cost)* — Actual cost to date for each of those categories.

- *Columns 3, 4 ($ Diff., % Diff.)* — The difference between estimated and actual cost (both in dollars and percent).

- *Column 5 (Act. Revenue)* — Actual revenue to date (the amount invoiced).

If you need information like this, you can use QuickBooks Pro to simplify collecting and recording job expenses so you get detailed reports from the beginning to the end of every job.

Why track job costs? Easy. There should be two profits in every job. The first is the money that ends up in your pocket. The second is what you learned. If you don't keep detailed cost records, you're missing part of the profit available on every contract. You're not finding the financial sinkholes that can turn any good job into a financial disaster. More likely, you'll keep making the same estimating mistakes over and over — until someone (your lender, your employees, your family) suggests that you get into another line of work.

What Is Job Cost Tracking?

Job cost tracking (sometimes called job costing) is one of QuickBooks Pro's major strengths. In fact, QuickBooks Pro is the only modestly-priced accounting program we know that does real job cost tracking. Every check you write for construction work — to suppliers, subcontractors, and employees — gets charged to a cost category in an estimate. Nothing sneaks by. You can see at a glance if costs are in line with the original bid.

Robins Construction
Job Estimates vs. Actuals Detail for Alan Palmer
All Transactions

	Est. Cost	Act. Cost	($) Diff.	(%) Diff.	Act. Revenue
Parts					
DemolitionMat	2,001.00	1,843.67	-157.33	-7.9%	2,001.00
ElectricalMat	20.00	21.00	1.00	5.0%	20.00
FlooringMat	1,240.00	1,196.47	-43.53	-3.5%	1,240.00
FoundationMat	1,890.13	2,046.56	156.43	8.3%	1,890.13
Int finishMat	1,668.50	3,217.95	1,549.45	92.9%	1,668.50
PaintingMat	966.24	897.42	-68.82	-7.1%	966.24
PlumbingMat	1,231.40	1,341.43	110.03	8.9%	1,231.40
Rough framMat	7,022.20	6,841.12	-181.08	-2.6%	7,022.20
Total Parts	16,039.47	17,405.62	1,366.15	8.5%	16,039.47
Service					
DemolitionLab	9,326.40	9,470.70	144.30	1.5%	9,326.40
FlooringLab	200.88	189.00	-11.88	-5.9%	200.88
FoundationLab	4,052.35	3,914.60	-137.75	-3.4%	4,052.35
Int finishLab	3,186.77	5,879.73	2,692.96	84.5%	3,186.77
PaintingLab	927.09	874.50	-52.59	-5.7%	927.09
PlumbingLab	724.90	943.00	218.10	30.1%	724.90
ProjectLab	1,250.00	1,250.00	0.00	0.0%	1,250.00
Rough framLab	5,301.94	4,940.00	-361.94	-6.8%	5,301.94
Total Service	24,970.33	27,461.53	2,491.20	10%	24,970.33
Other Charges					
ElectricSub	1,797.50	1,890.00	92.50	5.1%	1,797.50
FlooringSub	1,845.00	1,903.11	58.11	3.1%	1,845.00
FoundationEqu	1,282.56	1,100.00	-182.56	-14.2%	1,282.56
Overhead	4,593.49	4,500.00	-93.49	-2%	4,593.49
Profit	7,579.25	0.00	-7,579.25	-100.0%	7,579.25
Total Other Charges	17,097.80	9,393.11	-7,704.69	-45.1%	17,097.80
TOTAL	**58,107.60**	**54,260.26**	**-3,847.34**	**-6.6%**	**58,107.60**

Tracking costs on the Palmer job.

All the big engineering and construction companies have, and use, job cost tracking programs — usually costing thousands of dollars. Now you have one, too. If you haven't seen a job costing program in operation, keep reading. You may find this very interesting.

If your construction company estimates labor, material and equipment costs, and you have several crews working several jobs — or even one large job — you need to track job costs.

Here's an example. A contractor we know got a contract to build a large custom home. Most of the work was to be done with his crews. But 11 subcontractors would be required. Once work got started, invoices, purchase orders, checks, and bills were received and sent out nearly every day. This contractor's job cost system consisted of some numbers he scratched out with a stick on the dusty bed of his pickup truck. When material invoices for the foundation and framing came in, he paid the bills, not realizing that the cost was nearly 20 percent higher than quoted. A job cost tracking system would have identified the problem immediately. He should have called the vendors to complain. But he didn't do the math before paying the bills and didn't realize he'd been overcharged. Weeks later, when trying to figure out why he lost money on the project, he pulled out some old paperwork for the job and discovered the overcharges. By then it was too late.

If this example sounds familiar to you, it's time to start tracking costs on your jobs.

Estimates into Invoices

In Chapter 12 and Appendix A we mentioned that QuickBooks Pro estimating might not be the best choice for all contractors. The National Estimator program on the disk in the back of this book includes Job Cost Wizard, a bridge that exports completed estimates to QuickBooks Pro so you can use the accounting and tracking capabilities in QBP.

If you dream of turning estimates into invoices at the click of a mouse button, we'll show you how. If you need help, call the Craftsman Book Company support line at 760-438-7828 from 8 AM to 5 PM (Pacific) Monday through Friday. Help is free. All you pay for is the phone call.

We should let you know about one drawback to using our method to turn an estimate into an invoice. When you do this, you'll see that the material, labor, equipment, and subtotal costs in your estimate become billable items on the invoice. Maybe you don't want clients to see that much detail. Maybe you'd prefer to show just a lump sum cost for each job phase. We understand. Lump sum invoicing should be an option in QuickBooks Pro. Unfortunately, it's not. If you want to track labor, material, and subcontract expenses by job, you have to show labor, material, and subcontract expenses in the invoice. If you agree that tracking costs by phase (foundation, framing, finishing, etc.) and by type (labor, material, and subcontract) is valuable, you're probably ready to tolerate invoices that show too much detail. And there's another reason to identify material costs on your invoices — a little matter of sales tax.

Handling Tax

If you look back through the first 19 chapters of this book, you'll discover a significant omission. Tax doesn't appear anywhere. Why is that? Isn't tax important? Doesn't QuickBooks Pro handle tax? Are we suggesting that contractors should ignore sales tax?

Hardly! Unless all your work is for tax exempt government agencies, tax is an important cost in your jobs, usually 6 to 8 percent of material expense. But notice the word *usually*. Sales tax varies by state and province. And even within states, sales tax rates vary — not only by amount, but also by what gets taxed, when it gets taxed, and who has to collect and remit taxes to the taxing authority. Rules vary among the 50 states. And many states have dozens of taxing authorities, each with the right to set rates and decide what gets taxed and what doesn't. There's no uniform national rule so there's nothing we can say that applies everywhere.

However, QuickBooks Pro was developed for retail businesses to use, so it does an excellent job identifying what's taxable and what isn't, recording taxable sales, and even remitting the tax collected to the government authority.

True, some contractors can almost ignore sales tax. "Paper contractors" who subcontract everything in the job can ignore sales tax in most states. Connecticut is an exception. All of Canada is another exception. In Canada, contractors with sales volume over $30,000 have to add federal GST (Goods and Services Tax) to invoices and remit at least quarterly to the federal government. The only contractors in Canada that can ignore sales tax are contractors who work on government jobs exclusively.

Only a few states and provinces impose sales tax on labor. Only a few states and provinces require construction contractors to charge sales tax and remit the amount collected to a government agency. But nearly all construction contractors have to pay tax on at least some materials they purchase for installation. That makes tax an important cost component. So if you're going to pay tax on materials, you'll want to include tax in your material cost estimates. And therein lies the problem when we begin converting estimates to invoices.

How many of your material suppliers show shelf prices that include sales tax? For example, call for a quote on framing lumber. Does that quote include sales tax? The price posted on gasoline pumps includes tax. But we've never seen construction materials priced with tax. The tax is understood, not expressed. So, unless you fall under one of the exemptions listed above, you're going to pay sales tax on materials. Most general contractors learn very quickly to add sales tax on materials in their estimates. They have to.

If you use QuickBooks Pro to convert an estimate into an invoice, the invoice has to mirror sales tax in the estimate. If you pay tax on materials purchased for a job, you can mark each material subtotal on the invoice as taxable and accumulate a total for tax at the bottom of the invoice. Neat and easy! You pay tax on materials and that tax gets passed neatly and efficiently to your customers through invoices. But it only works if you identify material costs in your estimates.

Using Items for Job Cost Tracking

Look back at the three major cost types at the left margin in the Job Estimates vs. Actuals Detail report for the *Palmer* job.

- *Parts* includes all material costs
- *Service* includes all labor costs
- *Other Charges* includes equipment, subcontracts, overhead, and profit

Under each of the three cost types in the report, you see cost categories. For example, the first cost category under Parts is "DemolitionMat." This cost category covers demolition material expense such as tarps, tape, and hauling and dump fees. Demolition labor expense "DemolitionLab" is listed under the second cost type, "Service."

The first part of each cost category name comes from the estimate. The last three letters are either:

- *Mat* for material
- *Lab* for labor
- *Equ* for equipment
- *Sub* for subcontract

Cost categories in this estimate are Demolition, Electrical, Flooring, Foundation, Int finish, and so on. The estimator selected these descriptions when he created the estimate. It's OK to use any descriptions you want. You can use some list of "standard" construction phases or names you make up on the spot. Everything works. The 31 construction categories in *sample.qbw* or *company.qbw* are perfect if that's what you want. But they're not required.

Cost category names (the Item List) have to be fairly short. These short category names are fine when printing reports. Short names leave plenty of room on the page for cost columns. But it means you'll have to abbreviate category names at least occasionally when creating estimates.

Check the bottom line of the Palmer job report. The estimated cost is the same as the contract price, $58,107.60. The actual cost was $54,260.26. So the Palmer job was a winner by $3,847.34. Right? Look closer. Interior finish ran $4,242.41 over the estimate ($1,549.45 plus $2,692.96). Where does that money come from? You know already. Straight out of job profit. How long can you afford to wait to find problems like this?

One more question. What if you could get a job cost report like this (and lots more) on all your jobs without doing any extra paperwork, and using software you already have? What if there was a program that created the QuickBooks Pro Items for you automatically. Would you try it? Of course you would. And that's exactly what this appendix is going to explain.

National Estimator icon.

Job Cost Wizard icon.

The Job Estimates vs. Actuals Detail on page 303 is a QuickBooks Pro report. But the estimate we show in the figure was created in National Estimator, converted to QuickBooks Pro format by Job Cost Wizard, and then imported into QuickBooks Pro. You've got National Estimator and Job Cost Wizard on the CD in the back of this book. You also have over a hundred pages of manhour estimates and labor costs for residential and commercial construction.

If you estimate larger jobs and prefer to start with an existing labor and material cost estimating database rather than building your own Item List in QuickBooks Pro, this is the information you need.

Let's import an estimate we made using National Estimator and converted using Job Cost Wizard:

▌ In QuickBooks Pro, from the **File** menu, choose **Import**. Then click **IIF files** . . .

▌ Double-click **Home_off.iif**. When the estimate for this home office job has been imported, you may have to click **OK**. Ignore any warning message about job or estimate information.

▌ From the **Customers** menu, choose **Create Estimates**.

▌ Click **Previous** until you see the estimate for the Stillwel Job.

▌ Use the scroll bar at the right of the estimate to page through the estimate.

Importing an estimate (such as the Stillwel home office job) will set up all the items you need on the Items List. If you click the Create Invoice button in the Create Estimates window now, all costs in the Stillwel estimate will become charges in the Stillwel invoice. You probably won't want to do that.

To exit the Create Estimate window without saving the estimate:

▌ From the **Edit** menu, choose **Delete Estimate**.

▌ Click **OK**.

In this example, the Stillwel estimate had already been created in National Estimator and converted with Job Cost Wizard. But you'll want to create your own estimates. So the first step will be learning National Estimator.

Help Learning National Estimator

Show Me is an interactive video guide to National Estimator. Sit back and relax and let the 60-minute video run or jump from topic to topic. Exit any time you want. Then go back later to brush up on anything you missed. Show Me takes up very little space on your hard drive. But the Contractor's Guide to QuickBooks Pro 2005 disk has to be in your CD drive when watching Show Me.

There are three ways to start Show Me:

- You'll see links to Show Me on many Help topics when using the Help file.

- Click on **Show Me** in the Construction Estimating program group to run all of Show Me.

- Click on the "Show Me More" button below Julie's Tip of the Day to get more information about that tip.

Forty Estimating and Bidding Forms are included on Contractor's Guide to QuickBooks Pro CD. These forms are on the FORMS directory on the CD, and will install on your hard drive in My Documents\Construction Forms. If you have a laser or ink jet printer, you'll be able to create top-quality customized forms in minutes. To use these forms, you'll need any of the popular word processing or spreadsheet programs. First, start your word processing or spreadsheet program. Then click on **File** and **Open**. Change the drive letter to your CD (usually D:) and the directory to FORMS. You'll see over forty forms created especially for each of the following programs:

- Microsoft Word (file name extension of DOC)
- Microsoft Excel (file name extension of XLS)
- Lotus 1-2-3 (file name extension of WK4)
- Microsoft Works (file name extension of WKS)
- WordPerfect (file name extension of WP)

To open any form, double-click on the form name. Make the changes you want. Then save the modified form to your hard drive (such as C:). For a description of all forms on the CD, open the file INDEX.TXT on the FORMS directory.

National Estimator Help has everything you could ever want to know about National Estimator. These help files are available any time National Estimator

is running (click on the question mark) or from the Construction Estimating program group (click on **National Estimator Help**). To print a user's guide to National Estimator, click on the question mark. Then click on **Print All Topics**. Click on **File** on the Help menu bar. Click on **Print Topic**. Click on **OK**.

Job Cost Wizard Help explains how to use Job Cost Wizard to convert National Estimator estimates into invoices and then export the estimate or invoice to either QuickBooks or QuickBooks Pro (version 5 and higher). This help file is available any time you're running Job Cost Wizard. Just click **Help** on the menu bar. Then click on **Contents** and the topic of your choice. From Help Contents you can print a guide to Job Cost Wizard: Click on **Print All Topics**. Then click on **File**, **Print Topic** and **OK**.

System Requirements

National Estimator and Job Cost Wizard require Windows 98 or higher and about 100 Mb free on a hard drive.

Installing Contractor's Guide

With Windows running, put the CD in your CD drive (such as D:). If installation doesn't start after a few seconds:

1. Click on ![Start]
2. Click on **Settings**
3. Click on **Control Panel**
4. Double-click on **Add/Remove Programs**
5. Click on **Install**
6. Click on **Next**
7. Click on **Finish**

Follow the instructions on the screen. We recommend accepting the installation defaults. When installation is complete, click on Finish. National Estimator will begin automatically.

Uninstalling Contractor's Guide

Click on **Start**, **Settings**, **Control Panel**. Double-click on **Add/Remove Programs**. Then click on the name of the program to remove and **Add/Remove**.

Estimating with National Estimator

National Estimator begins when you click on the National Estimator icon or click on **Start**, **Programs**, the Construction Estimating group and then National Estimator.

On the title bar at the top of the screen you see the program name, National Estimator, and [Construction Estimating Reference Data]. Let's take a closer look at the other information at the top of your screen.

Appendix B: Job Cost Tracking and Importing Estimates

The Menu Bar

Below the title bar you see the menu bar. Every option in National Estimator is available on the menu bar. Click with your left mouse button on any item on the menu bar to open a list of available commands.

Buttons on the Toolbar

Below the menu bar you see 24 buttons that make up the toolbar. The options you use most in National Estimator are only a mouse click away on the toolbar.

Column Headings

Below the toolbar you'll see column headings for your costbook:

Craft@Hrs for craft (the crew doing the work) and manhours (to complete the task)

Unit for unit of measure, such as per linear foot or per square foot

Material for material cost

Labor for labor cost

Equipment for equipment cost

Total for the total of all cost columns

The Status Bar

The bottom line on your screen is the status bar. Here you'll find helpful information about the choices available. Notice "Page 20/371" near the center of the status line. That's a clue that you're looking at page 20 of a 371-page book. The book name is Construction Estimating Reference Data or "CERD" for short.

Check the status bar occasionally for helpful tips and explanations of what you see on screen.

The Costbook

All manhour estimating tables in Construction Estimating Reference Data are available in the Costbook Window. Notice also the words Page 20 Sitework at the left side of the screen just below the toolbar. That's your clue that the Sitework section of page 20 is on the screen.

Appendix B: Job Cost Tracking and Importing Estimates

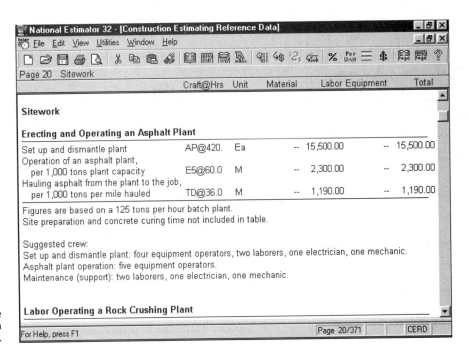

The Costbook Window has all the manhour estimates in Construction Estimating Reference Data.

To turn to the next page, either:

- Press [PgDn] (with Num Lock off), -or-
- Click on the lower half of the scroll bar at the right edge of the screen.

To move down one line at a time, either:

- Press the [↓] arrow key (with Num Lock off), -or-
- Click on the down arrow on the scroll bar at the lower right corner of the screen.

To turn quickly to any page, either:

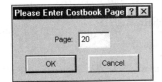

Type the page number you want to see.

- Click on the (Turn to Costbook Page) button near the right end of the toolbar, -or-
- Click on **View** on the menu bar. Then click on **Turn to Costbook Page**.

Type the number of the page you want to see and press [Enter↵]. National Estimator will turn to the top of the page you requested.

An Even Better Way

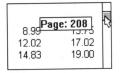

Drag the square to see any page.

Find the small square in the slide bar at the right side of the Costbook Window. Click and hold on that square while rolling the mouse up or down. Keep dragging the square until you see the page you want in the *Page:* box. Release the mouse button to turn to the top of that page.

Contractor's Guide to QuickBooks Pro **311**

Appendix B: Job Cost Tracking and Importing Estimates

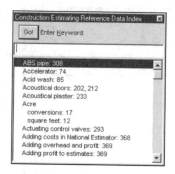

Use the electronic index to find cost estimates for any item.

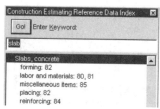

The index jumps to Slabs, concrete.

A Still Better Way: Keyword Search

To find any cost estimate in seconds, search by keyword in the index. To go to the index, either:

- Click on the ![Index] (Index) button near the center of the toolbar, -or-
- Click on **View** on the menu bar. Then press Enter↵.

Notice that the cursor is blinking in the Enter Keyword box at the right of the screen. Obviously, the index is ready to begin a search.

Your First Estimate

Suppose we're estimating the cost of a concrete slab on grade. Let's put the index to work with a search for slabs. In the box under Enter Keyword, type *slab*. The index jumps to the heading *Slabs, concrete*.

- Click once on the line *labor and materials: 80, 81* and press Enter↵, -or-
- Double-click on that line, -or-
- Press Tab and the ↓ arrow key to move the highlight to *labor and materials: 80, 81*. Then press Enter↵.

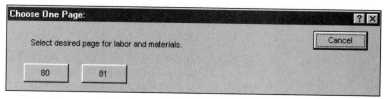

If costs appear on several pages, click on the page you prefer.

To select the page you want to see (page 81 in this case), either:

- Click on the number 81, -or-
- Press Tab to highlight 81. Then press Enter↵.

National Estimator turns to the top of page 81 as shown below.

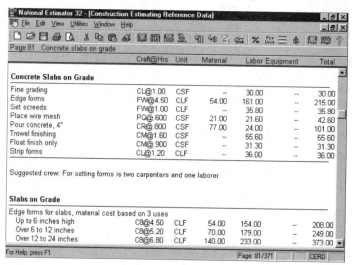

Costs for slabs on grade on page 81.

312 Contractor's Guide to QuickBooks Pro

Appendix B: Job Cost Tracking and Importing Estimates

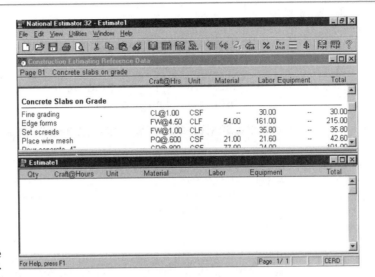

The Split Window: Costbook above and estimate below.

Splitting the Screen

Most of the time you'll want to see what's in both the costbook and your estimate. To split the screen into two halves, either:

- Click on the ▨ (Split Window) button near the center of the toolbar, -or-

- Click on **View** on the menu bar. Then click on Split Window and your screen should look like the example above.

The lines of the costbook on the top half of the screen are for slab base. The bottom half of the screen is reserved for your estimate. Column headings are at the top of the costbook and across the middle of the screen (for your estimate).

To Switch from Window to Window

- Click in the window of your choice, -or-

- Hold the [Ctrl] key down and press [Tab].

Notice that a window title bar turns dark when that window is selected. The selected window is where keystrokes appear as you type. Click on the bottom half of the screen so your estimate is selected.

Beginning an Estimate

You can type anything in the Estimate Window. Let's start by putting a heading on this estimate:

1. Press [Enter] once to space down one line.

2. Press [Tab] four times (or hold the space bar down) to move the Blinking Cursor (the insert point) near the middle of the line.

Contractor's Guide to QuickBooks Pro **313**

Appendix B: Job Cost Tracking and Importing Estimates

3. Type "Estimate One" and press Enter↵. That's the title of this estimate, "Estimate One."

4. Press Enter↵ again to move the cursor down a line. That opens up a little space below the title.

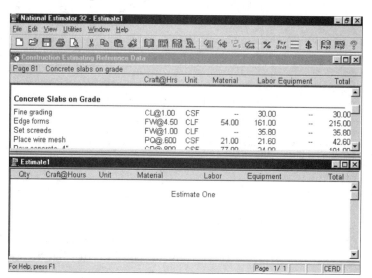

Begin by putting a title on your estimate, such as "Estimate One."

Copying Costs to Your Estimate

Next, we'll estimate the cost of 100 square feet of sand fill base. Click the (Split Window) button on the toolbar to be sure you're in the split window. Click anywhere in the costbook (the top half of your screen). Then press the ↑ or ↓ arrow key (with Num Lock off) until the cursor is on the line:

To copy this line to your estimate:

1. Click on the line.

2. Click the 📋 (Copy) button.

3. Click on the 📋 (Paste) button to open the Enter Cost Information dialog box.

Notice that the blinking cursor is in the Quantity box:

1. Type a quantity of 16 because this job requires 1600 square feet. (The unit of measure is CSF or 100 square feet.)

2. Press Tab⇥ and check the estimate for accuracy.

3. Notice that the column headed Unit Costs shows labor costs per unit, per CSF (100 square foot) in this case.

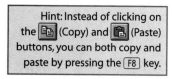

Hint: Instead of clicking on the 📋 (Copy) and 📋 (Paste) buttons, you can both copy and paste by pressing the F8 key.

314 Contractor's Guide to QuickBooks Pro

Appendix B: Job Cost Tracking and Importing Estimates

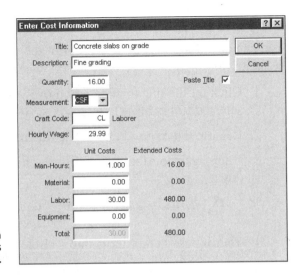

Use the Enter Cost Information dialog box to enter quantities and change costs.

4. The column headed Extended Costs shows costs for the entire 16 CSF (1600 square feet).

5. Press Enter⏎ or click on **OK** to copy these figures to the end of your estimate.

The new line at the bottom of your estimate shows:

Extended costs for fine grading as they appear on your estimate form.

16.00 is the quantity of fine grading

CL is the recommended crew, one laborer

@16.00 shows the manhours required for the work

CSF is the unit of measure, 100 square feet in this case

0.00 shows that no material cost has been entered in this case

480.00 is labor cost for the job

0.00 shows there is no equipment cost

480.00 is the total of material, labor and equipment columns

Copy Anything to Anywhere in Your Estimate

Anything in the costbook can be copied to your estimate. Just click on the line (or select the words) you want to copy and press the F8 key. It's copied to the last line of your estimating form. If your selection includes costs, you'll have a chance to enter the quantity. To copy to the *middle* of your estimate:

1. Select what you want to copy.

2. Click on the (Copy) button.

3. Click in the estimate where you want to paste.

4. Click on the (Paste) button.

Contractor's Guide to QuickBooks Pro **315**

Drag and Drop Estimating

You can also drag lines or words out of the costbook and drop them in your estimate:

1. Click on the line to copy and move your mouse slightly so the line is selected (turns black).

2. Release the mouse button. The line remains selected.

3. Click again on the selected line. This time hold your mouse button down.

4. Your mouse cursor turns into a circle with a diagonal bar.

5. Holding the mouse button down, move the circle with a bar into the Estimate Window.

6. Once in the Estimate Window, you'll see a vertical bar to the left of your mouse cursor.

7. Move this vertical bar to where the copied line should be pasted.

8. Then release the mouse button. The line is pasted in that position.

9. If the line pasted includes costs, you'll have a chance to enter a quantity.

Right-Click Editing

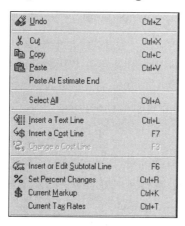

Right-click editing menu.

Most of what you do in National Estimator is editing — such as Cut, Copy, Paste and Undo. Most editing features have their own button on the toolbar. Of course, all editing functions are available from the Edit selection on the menu bar. But you might find it easier and quicker to open the floating edit menu by right-clicking with your mouse. Press the right mouse button. All edit functions currently a valid choice will be available. Simply click on the selection you need.

Appendix B: Job Cost Tracking and Importing Estimates

Changing Wage Rates

Search information on setting wage rates.

The labor cost for fine grading is based on a building laborer working at a cost of $29.99 per hour. Suppose $29.99 per hour isn't right for your estimate. What then? No problem! It's easy to use your own wage rate for any crew or even make up your own crew codes. To get more information on setting wage rates, press [F1]. At National Estimator Help, click on the **Key** button. Type *wage* then double-click on **Setting Hourly Wage Rates**. To return to your estimate, click on **File** on the National Estimator Help menu bar. Then click on **Exit**.

Changing Cost Estimates

With Num Lock off, use the [↑] or [↓] arrow key to move the cursor to the line you want to change (or click on that line). In this case, move to the line that begins with a quantity of 16. To open the Enter Cost Information Dialog box, either:

- Press [Enter ↵], -or-
- Click on the (Change Cost) button on the toolbar.

To make a change, either:

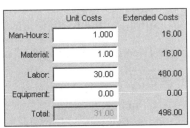

Change the material cost to 1.00.

- Click on what you want to change, -or-
- Press [Tab] until the cursor advances to what you want to change.

Then type the correct figure. In this case, change the material cost to $1.00.

Press [Tab] and check the Extended Costs column. If it looks OK, press [Enter ↵] and the change is made on your estimating form.

Changing Text (Descriptions)

Click on the (Estimate Window) button on the toolbar to be sure you're in the estimate. With Num Lock off, use the [↑] or [↓] arrow key or click the mouse button to put the cursor where you want to make a change. In this case, we're going to make a change on the line that begins "Concrete slabs."

To make a change, click where the change is needed. Then either:

- Press the [Del] or [←Bksp] key to erase what needs deleting, -or-
- Select what needs deleting and click on the (Cut) button on the toolbar.
- Type what needs to be added.

Contractor's Guide to QuickBooks Pro   **317**

Appendix B: Job Cost Tracking and Importing Estimates

Concrete **slabs** on grade
Fine grading
   16.00   CL@16.00

To select, click and hold the mouse button while dragging the mouse.

In this case, click just before "slabs." Then hold the left mouse button down and drag the mouse to the right until you've put a dark background behind "slabs." The dark background shows that this word is selected and ready for editing.

Press the [Del] key, or click on the (Cut) button on the toolbar, and the selection is cut from the estimate. If that's not what you wanted, click on the (Undo) button and "slabs" is back again.

Adding Text (Descriptions)

Some of your estimates will require descriptions (text) and costs that can't be found in Construction Estimating Reference Data. What then? With National Estimator it's easy to add descriptions and costs of your choice anywhere in the estimate. For practice, let's add an estimate for four reinforced corners to Estimate One.

Concrete slabs on grade
Fine grading
   16.00   CL@16.00
Reinforced corners

Adding "Reinforced corners."

Click on the (Estimate Window) button to be sure the estimate window is maximized. We can add lines anywhere on the estimate. But in this case, let's make the addition at the end. Press the [↓] arrow key to move the cursor down until it's just above the horizontal line that separates estimate detail lines from estimate totals. To open a blank line, either:

- Press [Enter ↵], -or-

- Click on the (Insert Text) button on the toolbar, -or-

- Click on **Edit** on the menu bar. Then click on **Insert a Text Line**.

Type *Reinforced corners* and press [Enter ↵].

Adding a Cost Estimate Line

Now let's add a cost for "Reinforced corners" to your estimate. Begin by opening the Enter Cost Information dialog box. Either:

- Click on the (Insert Cost) button on the toolbar, -or-

- Click on **Edit** on the menu bar. Then click on **Insert a Cost Line**.

1. The cursor is in the Quantity box. Type the number of units (4 in this case) and press [Tab].

2. The cursor moves to the next box, Measurement.

3. In the Measurement box, type *Each* and press [Tab].

4. Press [Tab] twice to leave the Craft Code blank and Hourly Wage at zero.

5. Since these corners will be installed by a subcontractor, there's no material, labor or equipment cost. So press [Tab] four times to skip over the Man-Hours, Material, Labor and Equipment boxes.

318 *Contractor's Guide to QuickBooks Pro*

Appendix B: Job Cost Tracking and Importing Estimates

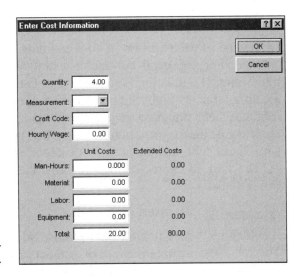

Unit and extended costs for four reinforced corners.

6. In the Total box, type 20.00. That's the cost per corner quoted by your supplier.

7. Press [Tab] once more to advance to OK.

8. Press [Enter] and the cost of four reinforced corners is written to your estimate.

Note: The sum of material, labor and equipment costs appears automatically in the Total box. If there's no cost entered in the Material, Labor or Equipment boxes (such as for a subcontracted item), you can enter any figure in the Total box.

Adding Lines to the Costbook

Add lines or make changes in the costbook the same way you add lines or make changes in an estimate. The additions and changes you make become part of the user costbook. For more information on user costbooks, press [F1]. Click on the key icon (Index). Type *add* and click on **Add lines to a costbook**.

Adding Tax

To include sales tax in your estimate:

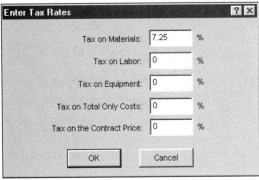

Type the tax rate that applies.

1. Click on **Edit**.

2. Click on **Current Tax Rates**.

3. Type the tax rate in the appropriate box.

4. Press [Tab] to advance to the next box.

5. Press [Enter] or click on **OK** when done.

In this case, the tax rate is 7.25% on materials only. Tax will appear at the end of the estimate.

Contractor's Guide to QuickBooks Pro **319**

Subtotals Become QuickBooks Cost Categories

At the end of each section in your estimate, you'll probably want to insert a subtotal. For a general contractor, estimate sections might be Demolition, Excavation, Foundation, Framing, etc. Estimate sections for an insurance repair contractor might include Kitchen, Bathroom or Living Room. Section subtotals help organize your estimates and make them easier to read and understand. Insert section subtotals wherever they make the most sense to you. These subtotals become cost categories when printing bids and invoices. The first 28 characters become cost category names when exporting to QuickBooks.

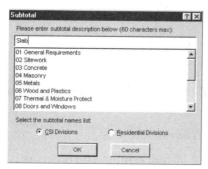
Insert a subtotal.

To insert a subtotal:

1. Click on the last cost line of the section (or on any blank line below the section).

2. Click on the (Subtotal) button on the toolbar (or click on **Edit** and **Insert Subtotal**).

3. Type a name or description for the section (such as *Slab*).

4. Press [Enter] or click **OK**.

If you prefer, select as a subtotal name one of the 16 CSI divisions or one of the 34 residential phases.

Adding Overhead and Profit

Set markup percentages in the Add for Overhead & Profit dialog box. To open the box, either:

- Click on the $ (Markup) button on the toolbar, -or-
- Click on **Edit** on the menu bar. Then click on **Markup**.

Type the percentages you want to add for overhead. For this estimate:

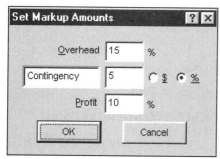
Adding overhead & profit.

1. Type *15* on the Overhead line.
2. Press [Tab] twice to advance to Contingency.
3. Type *5* on the Contingency line.
4. Press [Tab] twice to advance to Profit.
5. Type *10* on the Profit line.
6. Press [Enter] or click **OK**.

Markup percentages can be changed at any time. Just click on the $ (Markup) button and type the correct figure.

Appendix B: Job Cost Tracking and Importing Estimates

Preview Your Estimate

You can display an estimate on screen just the way it will look when printed on paper. To preview your estimate, either:

- Click on the 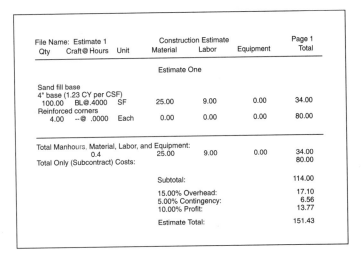 (Print Preview) button on the toolbar, -or-
- Click on **File** on the menu bar. Then click on **Print Preview**.

A preview of Estimate One.

In print preview:

- Click on **Next Page** or **Prev Page** to turn pages.
- Click on **Two Page** to see two estimate pages side by side.
- Click on **Zoom In** to get a closer look.
- Click on **Close** when you've seen enough.

Use buttons in Print Preview to see your estimate as it will look when printed.

Printing Your Estimate

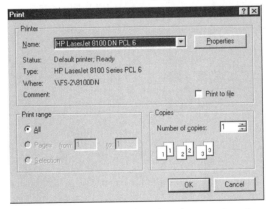

Options available depend on the printer you're using.

When you're ready to print the estimate, either:

- Click on the (Print) button on the toolbar, -or-
- Click on **File** on the menu bar. Then click on **Print**, -or-
- Hold the [Ctrl] key down and type the letter *P*. Press [Enter⏎] or click on **OK** to begin printing.

Contractor's Guide to QuickBooks Pro   **321**

Appendix B: Job Cost Tracking and Importing Estimates

Save Your Estimate to Disk

To store your estimate on the hard disk where it can be re-opened and changed at any time, either:

- Click on the 🖫 (Save) button on the toolbar, -or-
- Click on **File** on the menu bar. Then click on **Save**, -or-
- Hold the [Ctrl] key down and type the letter *S*.

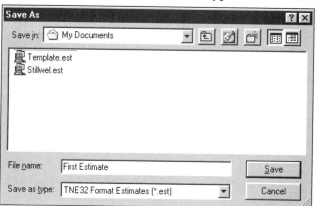

Type the estimate name in the File Name box to assign a file name.

The cursor is in the File name box. Type the name you want to give this estimate, such as *First Estimate*. Press [Enter ↵] or click on **Save** and the estimate is written to disk. Note that the default location for estimates is the *My Documents* folder.

Select Your Default Costbook

Your default costbook is the last costbook installed. It opens automatically every time you begin using National Estimator. Save time by choosing as the default costbook the one you use most.

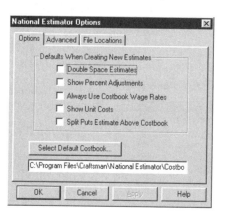

To change your default costbook, click on **Utilities** on the menu bar. Then click on **Options**. Next, click on **Select Default Costbook**. Click on the costbook of your choice. Click on **Open**. Then click on **OK**.

Selecting the default costbook.

Use National Estimator Help

Print a 37-page Guide to National Estimator.

That completes the basics of National Estimator. You've learned enough to complete most estimates. When you need more information about the fine points, use National Estimator Help. Click on the [?] (Help) button to see Help Contents. Then click on the menu selection of your choice. To print instructions for National Estimator, click on **Print All Topics** (at the bottom of the Welcome screen). Click on **File** on the Help menu bar. Then click on **Print Topic**. Click on **OK**.

Converting Estimates with Job Cost Wizard

Use Job Cost Wizard to:

- Convert estimates into bids and invoices you can send to a client, and,
- Export to QuickBooks where you can track job costs, receivables, payables, create payrolls and print financial reports.

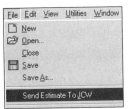
Send the estimate to Job Cost Wizard.

To view your completed estimate in Job Cost Wizard, either:

- Click on **File**, click on **Send Estimate to JCW**, or
- Hold the [Ctrl] key down and tap *J*.

You can also start Job Cost Wizard by clicking on the Job Cost Wizard icon in the Construction Estimating program group. Then click on the name of the estimate you want to open.

The Company Information dialog box will open the first time you use Job Cost Wizard. Type your company name and address. This will appear at the top of every estimate and invoice. When you've filled in infomation about your company, click on **OK** and your estimate will open in Job Cost Wizard.

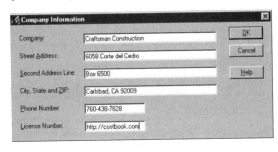
Fill in information about your company.

To change any of the information about your company:

- Click on **Options**.
- Click on **Your Company Info**.
- Type the changes needed and click **OK**.

Appendix B: Job Cost Tracking and Importing Estimates

Zoom, Scroll and Turn Pages

If the estimate doesn't fit your screen, set the percentage of zoom. For 640x480 resolution, type 81% in the zoom window and press ⌈Enter ⏎⌉.

Click and drag the vertical slide bar at the right of your screen to scroll down the page. Turn pages by clicking on the Previous Page or Next Page buttons.

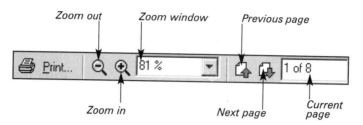

Enter Job Information

Job Cost Wizard needs some information about the job to create a nice-looking bid or invoice. For practice, enter job information for the Stillwel estimate:

1. **Customer name and address.** Click on the (Customer Info) button on the toolbar to enter information about the customer. Only the customer's first name and last name are required. All other information is optional. When done, click on **OK**.

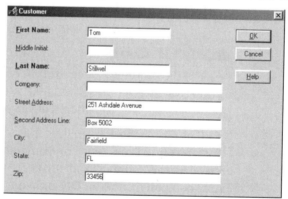

Fill in Customer Information.

2. **Job Name.** Click on the (Job Name) button on the toolbar. Then type the name of the job, such as *Room Addition*. When done, click on **OK**.

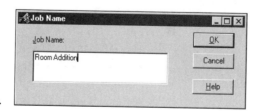

Enter a Job Name.

For transfers to QuickBooks — You can change the customer or job name after the file has been imported into QuickBooks. In QuickBooks, click on **Lists**. Click on **Customer:Job List**. Right-click on customer name or job name. Click on **Edit**. Then click on the tab of your choice.

3. **Estimate Number**. Click on the (Estimate Number) button on the toolbar. Job Cost Wizard keeps track of the last number used and recommends using the next number in sequence. Click on **OK** when done.

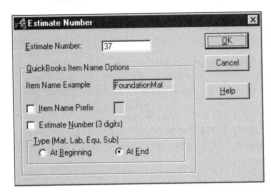

Check the Estimate Number.

For transfers to QuickBooks — When QuickBooks imports an estimate or invoice, subtotals in your estimate become cost categories "items" in QuickBooks. By default, cost category names in QuickBooks are the first 28 characters of estimate subtotal names plus the work type, either *Mat*, *Lab*, *Equ*, or *Sub*. You can change this default in the Estimate Number dialog box.

Job Cost Wizard Prints Invoices Your Way

Your estimates should cover every cost in a job. But your bids and invoices don't have to show all the details and reveal your markup. So Job Cost Wizard gives you choices about showing or hiding the details and markup.

Amount of Detail

To set the amount of detail, click on the (Details) button.

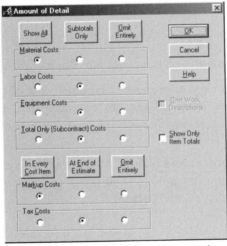

How much detail do you want to show?

If *Show All* is selected, every item in your bid or invoice will show a cost and each will become a cost category on the QuickBooks Item List. *Subtotals Only* is the default and will usually be a better choice.

If *Subtotals Only* is selected, subtotals will be the only costs in your bid or invoice. Each subtotal in your estimate becomes a cost category on the QuickBooks Items List. That's usually the best choice. If *Subtotals Only* is selected for all four cost categories, click on **Omit Work Descriptions** to show subtotal categories but hide all work descriptions.

The names you give to subtotals in National Estimator become cost category names in QuickBooks. Cost lines in National Estimator not followed by a subtotal become the "Project" subtotal.

If **Omit Entirely** is selected, neither costs nor descriptions will appear for that type of cost — either material, labor, equipment or total only (subcontract). Use Omit Entirely for materials, for example, when materials are being furnished by the owner.

"Total only" costs are assumed to be subcontract items. Subcontract items have a cost in the total column but no cost for material, labor, or equipment.

Markup and Tax

Use the three buttons at the bottom of the Amount of Detail dialog box to show or hide markup (overhead, contingency and profit) and tax.

In Every Cost Item distributes markup and tax proportionately throughout the estimate. There's no mention of overhead, profit or markup anywhere in the estimate or invoice.

At End of Estimate puts markup and tax at the end of the estimate, as in National Estimator.

Omit Entirely omits markup and tax from the estimate or invoice. Use this option if you prefer to add markup and tax in QuickBooks.

Click **Show Only Item Totals** if you don't want the invoice or estimate to show any breakdown of material, labor or equipment costs.

Click on **OK** when done with the Amount of Detail dialog box.

QuickBooks Account Names

Estimates and invoices imported into QuickBooks include expense and income account names. If the imported accounts do not exist already in your QuickBooks company, QuickBooks will create new accounts. You can control the names of these accounts by making changes in the QuickBooks Options dialog box.

- Click on the (QuickBooks Options) button on the toolbar, -or-

- Click on **Options** on the menu bar. Then click on **QuickBooks Options**.

Enter the names you prefer for income, cost of goods, markup and tax accounts. Change "Material Tax" to "FL Sales Tax," for example, if the job is taxable under Florida law. QuickBooks will keep track of tax due in each state where you do business.

Click **Use "Contractors Guide" Accounts** if you prefer the account names recommended in *Contractor's Guide to QuickBooks Pro*. To restore the default account names, click **Reset**.

Appendix B: Job Cost Tracking and Importing Estimates

If QuickBooks Pro version 2002 or later is installed on the computer, you should see a check mark beside *Use qbXML to integrate with QuickBooks*. XML exports to QuickBooks, as will be explained on the next page. If QuickBooks Pro is installed but there is no check mark, click on **Use qbXML to integrate with QuickBooks**. You'll be asked to identify the QuickBooks Company to receive imports from Job Cost Wizard. Select the company file you prefer and click **Open**. See *Exporting an Estimate to QuickBooks* on page 328 for more on opening the XML link to QuickBooks.

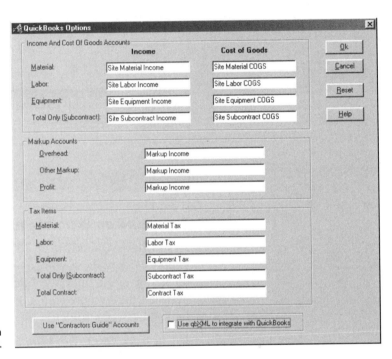

Click on OK when done with QuickBooks Options.

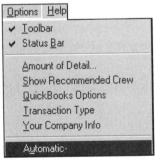

Set up for automatic operation.

Automatic

Job Cost Wizard always requires customer information, a job name and a job number before exporting an estimate. In automatic mode, Job Cost Wizard opens the Customer Info, Job Name and Estimate Number dialog boxes automatically after opening any estimate.

Job Cost Wizard runs in automatic mode when there is a check mark beside Automatic on the Options menu. To change to automatic mode, click on **Options** on the menu bar. Then click on **Automatic**.

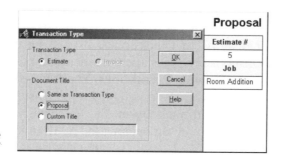

Changing the form title to "Proposal."

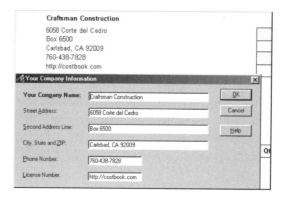

Enter your company name.

Transaction Type and Your Company Info

On the Options menu, click **Transaction Type** to change the form title from Estimate to Invoice to Proposal or anything you want (Custom). For QuickBooks exports, the transaction type has to be either estimate or invoice.

On the Options menu, click **Your Company Info** to change the company name or address.

Click on the **Print** icon or **File** and **Print**. Then click **OK** to print the document.

Exporting an Estimate to QuickBooks

If you have QuickBooks 2002 or later, Job Cost Wizard will select XML export by default. If you have an earlier version of QuickBooks, exports will create an Intuit Interface File (IIF). XML exports happen over a direct link between the two programs. For IIF exports, Job Cost Wizard writes a file in the QuickBooks folder.

If you have 2002 QuickBooks or later, use the XML export:

1. Begin by clicking the QB icon.
2. Job Cost Wizard will advise that QuickBooks is asking permission to access the company file.
3. If you receive a warning about security level, follow instructions on the screen.
4. Click **Launch QuickBooks** and select the company file to receive the export.
5. QuickBooks will open.
6. Click **Yes, Always** to grant Job Cost Wizard access to the QuickBooks company file.
7. A bar will report that the transfer is in progress. Click **OK** when the export is successful.
8. In QuickBooks, click **Customers**. Click **Create Estimates**. Then click **Previous** to see the imported estimate. You can skip information on IIF exports and go right to turning estimates into invoices at the bottom of page 330.

Appendix B: Job Cost Tracking and Importing Estimates

If you have an older version of QuickBooks, use the IIF export:

QB icon.

1. Begin by clicking the QB icon.
2. Change the drive or the folder if the QuickBooks folder is not listed at the right of Save in.
3. Check the file name to be sure it is what you want.
4. If you use QuickBooks Pro, click on **Estimate**.
5. If you use regular QuickBooks, click on **Invoice**.
6. When complete, click on **Save**.

Exporting an estimate to QuickBooks Pro does not affect the original estimate in any way. Job Cost Wizard can open and export an estimate as many times as you want. To create a second copy of the same estimate with different Job Cost Wizard options, save with a slightly different file name. But note that you can import an estimate for any customer and job only once. Details are on the next page.

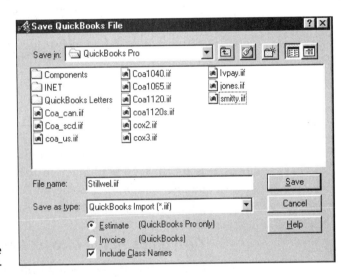
Exporting as an estimate to QuickBooks Pro.

Opening IIF Export Files in QuickBooks

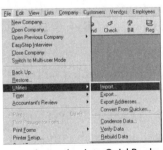
Importing into QuickBooks.

Once an estimate or invoice has been written to file with Job Cost Wizard, start QuickBooks:

1. Click on **File**.
2. Click on **Utilities**.
3. Click on **Import**.
4. Double-click on the name of the estimate or invoice you want.
5. Click on **OK** when the import is complete.
6. Click on **Customers**.
7. Click on **Create Estimates** if you saved the file as an estimate.
8. Click on **Create Invoices** if you saved the file as an invoice.
9. Click on **Previous** to see the file just imported.

Important Note: With QuickBooks Pro 2000 and higher, you can import an estimate only once from an IIF. On second import of the same estimate, you'll see an error message, "Can't record invalid transaction." If you make a mistake and want to import an estimate again, delete the previous imported estimate before importing again. Instructions for deleting an estimate begin under *Don't Worry About Making a Mistake*. If you want two versions of any one estimate in QuickBooks Pro, save the alternate estimate with a slightly different job name or customer name. That makes the estimate different enough so it will import perfectly into QuickBooks.

Filling in The "Amount" Column (IIF only)

An imported estimate is not complete until some figure appears in the Amount column of each cost line. So long as the Amount column is blank, QuickBooks will consider the estimated cost for that line to be zero (even when numbers appear in the Total column). If the Amount column for an entire estimate is left blank, QuickBooks reports will show the estimated cost for that job to be zero.

Cost	Amount	Markup	Total
3,544.38	3,544.38		3,544.38
,924.38			3,924.38
			7,468.76

Forcing a figure into the "Amount" column.

The fastest way to fill in the Amount column for a QuickBooks estimate is to click on a number in the "Cost" column and change the figure by a penny. Continue clicking and changing costs by a penny until every row of costs includes a figure in the Amount column.

Turn an Estimate into an Invoice in QuickBooks

First, decide if you want an invoice for the whole job or for just part of the job (progress billing). If you prefer progress billing:

1. Click on **Edit**, click on **Preferences**, click on **Jobs & Estimates**, click on **Company Preferences**, click on **Yes** under **Do You Do Progress Invoicing?**. Click on **OK**.

To create the invoice:

1. Click on the **Create Invoice** button at the top of the QuickBooks estimate screen.
2. Click on **Yes** to record changes to the estimate.
3. If you selected progress billing, enter a percentage or select items to be invoiced.
4. Click on **OK** and QuickBooks creates the invoice.
5. Make changes to the invoice if you want.

6. Click to print one copy for your file and another for your customer.

7. Click on **OK** when done. (Processing the file may take a little time.) QuickBooks reports will now include totals from the job just invoiced.

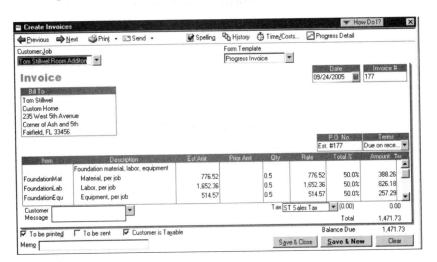

An invoice for the first half of the foundation work.

Don't Worry About Making a Mistake

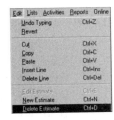

It's easy to delete an estimate.

QuickBooks is very forgiving. Practice all you want. Experiment any way you want. Then delete any estimate or invoice to remove every trace of it from QuickBooks. With the offending estimate or invoice displayed:

1. Click on **Edit**.
2. Click on **Delete Estimate** (or Invoice).
3. Click on **OK**.
4. The estimate (or invoice) is deleted.

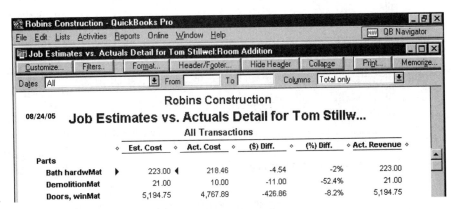

Estimates vs. Actuals Detail.

Appendix B: Job Cost Tracking and Importing Estimates

Your Jobs in QuickBooks

- Click on **Reports**, **Company & Financial**, **Profit & Loss By Job**, to see job income and expense.

- Click on **Reports**, **Company & Financial**, **Balance Sheet Standard** to see the new receivables total.

- Click on **Reports**, **Jobs & Time**, **Job Estimates vs. Actuals Detail**, select the customer and job to see a detailed cost comparison for the job. Until you start paying bills, the Actual Cost column will be all zeros.

Paying Bills by Cost Category

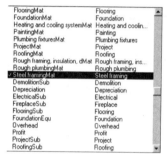

Select a cost category from the Item list.

When you pay vendors and subcontractors the amount paid is charged to the job and deducted from your bank balance.

1. Click on **Banking**.
2. Click on **Write Checks**.
3. Fill in the **Pay to the Order** line.
4. Click on the **Items** tab.
5. Click on the down triangle in the Item column of the check stub to open the Item list.

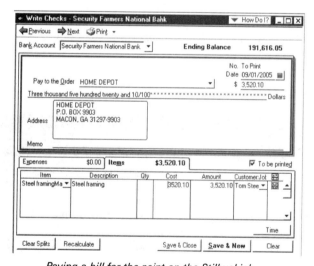

Paying a bill for the paint on the Stillwel job.

6. Select a *Mat*, *Sub* or *Equ* cost category from the Item list. These are the material, subcontract and equipment subtotal costs from your estimate.

7. Click in the amount column and enter the amount paid for the item described. Click on the down triangle under Customer:Job and select the correct customer and job.

8. This amount is not billable to your customer. So click on the icon representing an invoice to put a red X over the icon.

9. One check can cover items in several cost categories and even costs on several jobs. Click again on the next line down in the Item column and find the next cost item.

Creating Payroll by Cost Category

When you write payroll checks, the amount paid is charged to the job and deducted from your bank balance.

1. Click on **Employees**.

Appendix B: Job Cost Tracking and Importing Estimates

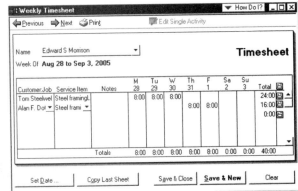

Fill out the timesheet by customer and service item.

2. Click on **Time Tracking**.

3. Click on **Use Weekly Time Sheets**.

4. Click on the down triangle opposite Name and select the employee to be paid.

5. On the Timesheet, click on the down arrow under Customer:Job and select the first job where the employee worked.

Under Service Item, select a labor cost category from the list.

6. Under Service Item, select a Lab cost category from the list. These are labor subtotal costs from your estimate.

7. Under Payroll Item, select the type of pay, such as hourly.

8. Enter the number of hours worked for that cost category.

9. Any timesheet can cover work done on several jobs and many service items.

10. When finished recording time for an employee, click on **Save & Close**.

To actually produce paychecks:

1. Click on **Employees** and then **Payroll**.

2. Click on **Pay Employees**.

3. Check the names of the employees to be paid.

4. When done, click on **Create**.

Index

A

A/P Aging Detail report 253-254, 259
A/R Aging Detail report 250-252, 259
Accounting, setting QuickBooks Pro
   preferences 34-36
Accounts payable 175-190
   A/P Aging Detail report 253-254, 259
   accrual basis 175-190
   bills 175-190
   checking workers' comp expiration
      . . . . .186-187, 233-234
   entering opening balance bills . . .132-133
   purchase orders 176-179
Accounts receivable 151-173
   A/R Aging Detail report 250-252, 259
   accrual basis 151
   entering opening balance invoices
     130-132
Accounts, making inactive 25, 31
Accounts, organizing with
   QuickBooks Pro 6
Accrual basis 151, 175-190
   payables 175-190
   reasons for using 205
   receivables .151-173
   setting QuickBooks Pro report
      preferences 46-47
Additional info tab 123-124
Address and Contact tab 123
Adjustment register 261
Aging detail reports 259
Assets
   Chart of Accounts 55-57
   Chart of Accounts, sample 57-62
   current 55
   fixed 55-56
Assisted Payroll 195
Audit trail, setting QuickBooks Pro
   preferences 35
Author biographies 9-10

B

Backup
   annual reports 259-260
   end of month reports 254-255
Bad debts, writing off 259
Balance sheets
   comparison 255-258
   end of month 254
   end of year 259
   printing 130
Balances, opening 129-130
Basic Payroll Service 195-196
Basis, accounting
   accrual 205
   cash 205-211
   setting QuickBooks Pro report
      preferences 46-47, 206
Bidding forms, CD 308
Bills
   entering opening balances for
      Accounts Payable 132-133
   entering without purchase orders
     181-182
   job related expenses 181-182
   overhead expenses 182-183
   paying by cost category 332-333
   selecting for payment 183-185
   Unpaid, by job .246
Biographies, authors 9-10
Builder's Guide to Accounting 286
Business Planning .73

C

Capital, Chart of Accounts 55-56
Cash basis accounting 205-211
   checking transactions 211
   reasons for using 205
   recording checks 206-209
   recording deposits 209-210
   recording job related costs 207-208
   recording overhead costs 208-209
   setting QuickBooks Pro preferences . .206
   setting QuickBooks Pro report
      preferences 46-47
Cash Flow Projector 72
CD-ROM
   contents .12, 19-20
   installing .13-14, 17
   Show Me demonstration 12, 17
   system requirements 13
   uninstalling 14
CD-ROM, Contractor's Guide 308
CERD (Construction Estimating
   Reference Data) 310-312
Change orders, tracking 162-164
   estimate, reviewing and changing 164
Chart of Accounts
   adding accounts 63-65, 261
   adjustment register 261
   assets account 55-56
   capital account 55-56
   changing accounts 63-65, 261
   company.qbw 57-62
   construction loan account 263
   corporation 56
   escrow account 262, 270
   income account 55-57
   job costs account 55, 57
   land acquisition loan 261, 270
   land development loan 275
   land purchase .262
   liabilities account 55-56
   linking accounts to items 75
   loans payable 267-270
   opening balances, entering 129-130
   overhead costs account 55, 57
   partnership 56
   printing 63-65
   report 218-219
   revenue account 55-57
   sample Chart of Accounts 57-62
   sample.qbw 57-62, 85
   setting up retentions account . . .164-167
   sole proprietorship 56
   subaccounts 64
   Work in Process account 262
Check runs 183-185
Checking account, reconciling 247-254
Checking, setting QuickBooks Pro
   preferences 36-37
Checks, printing 184-185
Class tracking, setting QuickBooks
   Pro preferences 34-35, 104
Classes 103-105
   setting Class Tracking preference
      . . . . .34-35, 104
   subclasses 105
   tracking cost categories 103
   tracking job costs 103
   updating 29
Closing costs, land purchase 266-269
Closing, end of month
   backup 254-255
   checking transactions 249-254
   procedures .247-255
   profit and loss statements 254
   reports 249-254
   reports to print 254
Closing, end of year
   backup 260

Index

financial statements 259-260
procedures 259-260
profit and loss statements 259
reports to print 259
Committed costs, tracking 177-180
Company contributions, workers'
  comp, setting up payroll items 89-93
Company data file
  condensing 299
  converting from Quicken 25-30
  locating on hard drive 18
  renaming 26-27
  upgrading older version 23-25
Company files 18
Company information, adding &
  changing 21-22, 27
company.qbw 144
  changing name 22
  Chart of Accounts 57-62
  Chart of Accounts report 218-219
  construction categories 306
  contents 20
  Item List in 75
  Job Cost Class report 220-222
  Labor Cost by Job report 224-226
  locating 20-21
  memorized reports 217-234
  opening 20-21
  renaming 22
  Unpaid Job Bills report 229-232
  Vendor Comp Expirations report
 233-234
  Work in Process account 262
  Workers' Compensation report ...226-227
Complete Payroll Service 196
Computerized accounting 286
Construction business, organizing
  with QuickBooks Pro 7-8
Construction draws 279-281
Construction estimating programs
 301-302
Construction Estimating Reference
  Data (CERD) 310-312
Construction Loan account 263
Contractor's Guide CD-ROM
  contents 12, 19-20
  estimating and bidding forms 308
  installing 13-14, 17, 309
  system requirements 13, 309
  uninstalling 14, 309
Converting existing data file 30-31
Corporation, Chart of Accounts 56
Craftsman Book Company
  tech support 14, 302, 304
  Web site 286, 302
Create an estimate 293-294
Current Liability accounts 66
Current tax tables, obtaining 12
Custom field, adding to purchase order
 188-189
Customer:Job List 107-111
  adding a customer 107-111

exporting 114
printing 114
updating 28
Customers 107-113
  adding job for 111-113
  change orders 162-164
  fixed price contract invoices 157-159
  invoices 152-162
  organizing files for 136-137
  recording payments 169-171
  set price invoices 152-154
  setting QuickBooks Pro preferences ...48
  setting up 107-111
  time-and-materials or cost-plus
    invoices 155-157
  transactions flow chart 137
  updating 28
Customizing, estimate forms 146-147

D

Demonstration, CD-ROM 12, 17, 308
Deposit item 81-82
Deposits
  construction draws 279-281
  printing customer receipts for 172
  recording 170-173
  recording job deposits 171-173
  recording on a cash basis 209-210
Developers
  construction draws 279-281
  deposits 281-282
  payments 281-282
  recording sales 282-285
  setting up development job 263-264
  using items 263-265
Development loans 275-279
  closing escrow 275-277
  costs 278
  recording payoff 278
Development, real estate
  construction draws 279-281
  deposits 281-282
  payments 281-282
  recording sales 282-285
  setting up development job 263-264
  using items 263-265
Disk, CD-ROM
  contents 12, 19-20
  installing 13-14, 17
  system requirements 13
  uninstalling 14

E

Easy Step Interview 18
Employee Payroll Info Tab
  ....123-127, 193-194
Employees
  adding workers' comp cost to jobs ...199
  billing labor on time & materials
    invoices 194

correcting timesheets 198
entering timesheets 191-194
missing payroll items on timesheet ...198
organizing files for 139
paycheck additions 199
paycheck deductions 199
payroll info tab 123-127
setting QuickBooks Pro preferences
 42-43
setting up custom fields for 123-124
setting up new employees 121-128
template 43
End of month procedures 247-255
backup 254-255
checking transactions at month's
  end 249-254
End of year procedures 259-260
financial statements 259-260
profit and loss statements 259-260
reports 259-260
Retained Earnings account 35-36
writing off bad debts 259
Escrow, land development
  closing 266-269
  deposit account 262
  earnest money 265-266
  recording fees 278
  sample statement 276
Estimates, creating with
  National Estimator 309-323
  adding cost estimates lines 318-319
  adding descriptions 318
  adding lines to costbook 319
  adding overhead and profit 320
  adding tax 319
  changing cost estimates 317
  changing descriptions 317-318
  changing wage rates 317
  column heading 310
  converting using Job Cost Wizard
   323-331
  copying anything 315
  copying costs from 314-315
  costbook 310-314
  default costbook 322
  drag and drop estimating 316
  editing features 316
  getting help with 14, 308, 323
  importing into QuickBooks Pro 307
  installing 308-309
  invoices, making from 304
  keyword search 312
  menu bar 310
  previewing estimates 321
  printing estimates 321
  saving estimates 322
  splitting screen 313
  status bar 310
  subtotals in 320, 325
  switching windows 313

Index

tech support 14, 302, 304
toolbar 310
Estimates, creating with QuickBooks Pro
  adding a tax due item 296
  advantages of 287-288
  construction estimating
    programs 301-302
  creating 293-296
  customizing estimate form 146-147
  deleting 331
  grouping items 295
  handling sales taxes 304-305
  importing files to QuickBooks Pro 307
  Job Cost Wizard 149
  making invoices from 296-298, 304
  markup 146, 326
  memorizing 147
  National Estimator 149
  new items .293-294
  progress billing 149
  reviewing and changing 164
  setting QuickBooks Pro items for 144
  setting QuickBooks Pro preferences
    . . . . .41, 289-290
  setting up Item List 289
  specialty contracting 288-299
  taxables 289-290
  tracking job costs 302-304
  tracking sales tax 296
  transferring summary estimate
   144-146
  using memorized estimate 148
  using QuickBooks Pro 288-299
Estimates, turning into invoices 304
Estimating and Bidding forms, CD 308
Estimating programs 301-302
Estimating programs included 10
Examples, folder .21
Expenses, Chart of Accounts 55, 57
Expert Analysis Tool 73

F

Fees, land development 278
File name for New Company 22
Files, office
  keeping current 139-141
  organizing 136-141
Filters tab, modifying reports 216
Finance charges, setting QuickBooks Pro
  preferences .38-40
Financial statements
  analyzing .255-258
  balance sheets 255-256, 259-260
  end of month .254
  end of year 259-260
  printing 254, 259
  profit and loss statements
    .255-256, 259-260
Fixed Asset Item List 65-66
Fixed Asset Tracking 65-66
Fixed price contract invoices 157-159

Forecasting 72
Forms
  estimating, CD 308
  IRS 1099 forms 52-53
  IRS 1099 forms for subcontractors
   115-119
  IRS 1099 forms for vendors 115-119
  send forms preference 48-49
  setting QuickBooks Pro preferences
   52-53
  setting up non-1099 subcontractors
   119

G

GAAP 8
Generally Accepted Accounting
  Principles 8
Getting started with QuickBooks Pro 17
Graphs, setting QuickBooks Pro
  preferences .46-47
Group of items 80-81

H

Help
  how to contact Craftsman Books
    . . . . .14, 286, 302, 304
  how to contact Intuit 9, 14, 195
  how to contact Online Accounting
    . . . . .9, 14, 286
  Job Cost Wizard 14, 309, 323-333
  National Estimator 14, 308, 323

I

Icons
  Job Cost Wizard 307
  National Estimator 307
Import files, opening in QuickBooks
  Pro 307
Income, Chart of Accounts accounts
 55-56
Installing, Contractor's Guide
  CD-ROM 13-14, 17, 309
Integrated Applications, setting
  QuickBooks Pro preferences 40
Interest payment, personal loan . . .272-273
Intuit
  phone number 9, 14, 195
  QuickBooks Pro trial version 8-9, 17
  technical support 14
  Web site 8, 17, 286
Inventory
  items 78-80
  setting QuickBooks Pro preferences
   43-45
Inventory based business, items 78-80
Invoices
  applying deposit to 173
  creating from estimates 304, 329
  creating with Job Cost Wizard . . .325-328
  deleting 331
  entering opening balance for
    Accounts Receivable 130-132

fixed price contract 157-159
  including sales taxes 305
  making from estimates 296-297, 304
  progress invoices 41, 159-162
  recording payments 169-171
  recording sale of property 282-285
  retainage 164-168
  set price 152-154
  terms of 153-154
  time and materials 155-157
  tracking through QuickBooks Pro 298
  when to delete 298-299
IRS 1099 forms
  for subcontractors 115-119
  for vendors 115-119
  setting QuickBooks Pro preferences
   52-53
Item List window 76
Item list, planning 288, 291-293
Items 75-84
  adding tax due into estimate 296
  deposit 81-82
  description limits 288
  grouping 80-81, 295
  inventory based business 78-80
  linking to Chart of Accounts 75
  loan fees and taxes 278
  making inactive 25, 31
  new for estimates 293-294
  non-inventory based business 76-78
  non-job related items 81-84
  payroll 85-102
  real estate development 263-265
  setting up retentions 164-165
  subitems 77-78
  updating 29
  using to track job costs 306-307

J

Job Cost Class report 220-222
Job Cost Wizard 10, 304
  automatic mode 328-329
  creating invoices 325-328
  description of 12, 149
  entering job information 324-325
  entering QuickBooks account
    namesyt44 326
  fitting to screen 324
  help 14, 309
  icon 307
  installing 308-309
  markup 326
  starting 323
  tech support 14, 304
  using to convert estimates 323-331
Job costs
  Chart of Accounts 55, 57
  tracking owner's time 127-128
  tracking through QuickBooks Pro
   302-307
  using items to track 306-307

Index

Job Costs Detail report 227-229
Job deposits, recording 171-173
Job Estimates vs. Actuals
   Detail report 238, 303, 307
   Summary report 237-238
Job Estimates vs. Actuals Detail 143
Job Profitability Detail report 236-237
Job Profitability Summary report 235
Job Progress Invoices vs. Estimates
   report 239-240
Job Report, unbilled costs 240-241
Job status 112-114
Job transactions, flow chart 137
Job types 112-113
   subtypes 112-113
Job-related expenses, entering bills
  181-182
Jobs 28
   adding 111-113
   adding workers' comp cost to jobs
  199-200
   change orders .141
   Chart of Accounts costing account
   . . . . .55, 57
   IRS 1099 forms for subcontractors
  115-119
   IRS 1099 forms for vendors 115-119
   organizing files for 136-137
   recording job deposits 171-173
   setting QuickBooks Pro preferences . . .41
   status 112-114
   subtypes 112-113
   tracking costs through QuickBooks
   Pro 302-307
   transactions flow chart 137
   types 112-113
   using items to track costs 306-307
Jobs, Time & Mileage project reports . . .234
Julie, Show Me demonstration 17

L

Labor Cost by Job report 224-226
Labor costs, grouping 295
Land acquisition loan, Chart of
   Accounts account 261, 270
Land development 261-286
   interest, closing costs 266-269
   loan transactions 269-270
   loan, Chart of Accounts account 275
   recording fees and taxes 278
Land purchase transactions 265-266
   closing escrow 266-269
   personal loan 269-270
   seller carry back loan 270
Land purchase, Chart of Accounts
   account 262
Liabilities, Chart of Accounts account
  55-56
Loan Manager 66-72
   example 66-72
   making loan payments from 72
   window 71

Loans payable 262, 269
   Chart of Accounts account 269-270
Long Term Liability accounts 66

M

Making items inactive 25, 31
Manual Calculations (payroll) 196
Markup
   imported estimates, in 326
   Job Cost Wizard 326
   QuickBooks Pro estimate 146, 293-294
Material costs, grouping 295
Memorized reports 217-234
   Chart of Accounts 218-219
   Job Cost Class 220-222
   Job Costs Detail 227-229
   Labor Cost by Job 224-226
   Unpaid Job Bills 229-232
   Vendor Comp Expirations 233-234
   Workers' Compensation 226-227
Memorizing, estimates 147-148
Modifying reports 213-216

N

Name limits, items 288
National Estimator 289, 301
   adding cost estimates lines 318-319
   adding descriptions 318
   adding lines to costbook 319
   adding overhead and profit 320
   adding tax 319
   changing cost estimates 317
   changing descriptions 317-318
   changing wage rates 317
   column heading 310
   copying anything 315
   copying costs from 314-315
   costbook 310-314
   creating estimates with 309-323
   default costbook 322
   description of 12
   drag and drop estimating 316
   editing features 316
   getting help with 14, 308, 323
   imported estimate 10
   keyword search 312
   menu bar 310
   previewing estimates 321
   printing estimates 321
   saving estimates 322
   splitting screen 313
   status bar 310
   subtotals in 320, 325
   switching windows 313
   tech support 14, 304
   technical support phone line 304
   toolbar 310
   tutorial 12, 308
National Estimator Help 308
New Item window .77

Non-inventory based business 76-78
Non-job related items 81-84

O

Office files
   keeping current 139-141
   organizing .136-141
Office organization 135-141
   keeping current 139-141
Office work flow, organizing 135-141
Online Accounting, Web site 9, 286
Open Purchase Orders by Job Report
  241-242
Opening balances
   entering in Chart of Accounts 129-130
   entering in QuickBooks Pro 129-130
   entering open balance bills for
   Accounts Payable 132-133
   entering open balance invoices for
   Accounts Receivable 129-132
   printing balance sheet 130
Organizing work flow 135-141
Overhead
   Chart of Accounts account 55, 57
   entering in estimates 326
   Job Cost Wizard 326
   National Estimator estimates 320
   QuickBooks Pro estimate 146
Owner's timesheet 201-204

P

Paperwork
   change orders .141
   customer files 136-137
   employee files .139
   job files 136-137
   keeping current 139-141
   organizing .136-141
   payroll files 139
   timesheets .139-140
   vendor files 138
Partnership
   adding partner's costs to jobs 201
   Chart of Accounts 56
   setting up payroll items 96-102
   timesheets .201-204
Payables 10, 175-190
Paychecks 195-200
   printing 200
Payments
   construction draw, entering 279 281
   recording customer payments . . .169-171
   selecting bills for payment 183-185
Payroll info tab 123-127, 193-194
Payroll items 85-102
   adding 93-95
   company workers' comp
   contributions 85-93
   deduction for owner 96-102
   deduction for partner 96-102
   for not tracking workers' comp 93-95

Index

linking to Chart of Accounts 85-102
owner or partner 96-102
QuickBooks Pro defaults 93
tracking workers' comp costs 85-93
updating 29
workers' comp classifications 86-89
Payroll service, QuickBooks Enhanced ...43
Payroll, See also Payroll items
................................11, 191-194
adding workers' comp cost to jobs
.................................199-200
correcting timesheets 198
entering items not set up 193-194
entering timesheets 191-194
missing payroll items on timesheet ...198
organizing files for 139
paycheck additions 199
paycheck deductions 199
processing 195-197
processing for owners 202-204
processing for partners 202-204
services 195-196
setting QuickBooks Pro preferences
................................42-43
Personal loans 269-275
recording deposit 271-272
recording interest/principal 272-275
recording repayment 274-275
Personal tab 122
Phone numbers
Craftsman Book Company ...14, 302, 304
Intuit 9, 14
Preferences 33-54
accounting 34-36
audit trail 35
checking 36-38
class tracking 104
customers 48
employee template 43
employees 42-43
estimates 41
finance charges 38-40
general 33
graphs 46-47
integrated applications 40
inventory 43-45
IRS 1099 forms 52-53
jobs 41
payroll 42-43
progress invoicing 41
purchase orders 43-45
reminders 45
reports 46-47
sales 48-49
sales tax 48-49
send forms 48-49
service connection 50
setting for estimating 289-290
spelling 51
time tracking 54
vendors 43-45

Preview Paycheck 198
Principal payment, personal loan ..274-275
Printing
balance sheets 130
Chart of Accounts 63-65
checks 184-185
customer receipts for deposits 172
customers 114
financial statements 254, 259
National Estimator estimates 321
paychecks 200
profit and loss statements 254, 259
receipts 172
Profit
imported estimates, in 326
Job Cost Wizard 326
National Estimator estimates 320
QuickBooks Pro estimates ...146, 293-294
Profit and loss statements 255-258
end of month 254
end of year 259-260
printing 254, 259
Progress billing 149, 289
Progress invoice 41, 159-162
Progress invoicing setting QuickBooks
Pro preferences 41
Project reports 234-246
Job Estimates vs. Actuals Detail
..............................238, 303, 307
Job Estimates vs. Actuals Summary
..................................237-238
Job Profitability Detail 236-237
Job Profitability Summary 235
Time by Job 241
Purchase orders 11
adding custom fields 188-189
checking workers' comp expiration
..................................186-187
creating 176-177
creating custom fields 188-189
setting QuickBooks Pro preferences
................................43-45
tracking committed costs 177-180
tracking multiple draws 177-179
tracking unfilled orders 180-181
using custom fields 188-190
using to track draws and buyouts
..................................177-179

Q

QuickBooks Enhanced Payroll 195
service 43
QuickBooks Premier
Accountant Edition features 65
Contractor Edition features 65
Contractor Edition, using reports in
..................................244-246
QuickBooks Pro
advantages 7-8
construction features 10
estimating with 288-299

getting a trial version 8-9, 17
getting started 17
organizing accounts with 6
technical support 14
QuickBooks programs, choosing 7-8
QuickBooks Standard Payroll 195
Quicken, converting files from 25-30

R

Real estate development 261-286
closing escrow 266-269
construction draws 279-281
deposits 281-282
escrow deposits 262
land purchase transactions 265
payments 281-282
personal loans 269-275
recording fees and taxes 278
recording sales 282-285
seller carry back loan 270
setting up development job 263-264
using items 263-265
Receivables 10, 151-173
accrual basis 151-173
entering opening balance invoices
..................................130-132
setting up retentions account ...164-168
Reconciling, checking account 247-254
Recording
escrow fees, land development 278
invoice, sale of property 283-284
sales 282-285
Reminders, setting QuickBooks Pro
preferences 45
Renaming *company.qbw* file 22
Reports 213-246
A/P Aging Detail 253-254, 259
A/R Aging Detail 250-252, 259
balance sheets 255-258, 259
cash flow projector 72
Chart of Accounts 218-219
Cost to Complete by Job Detail ..244-245
end of month 249-254
end of year 259-260
Job Cost Class 220-222
Job Costs Detail 227-229
Job Estimates vs. Actuals Detail
..............................238, 303, 307
Job Estimates vs. Actuals Summary
..................................237-238
Job Profitability Detail 236-237
Job Profitability Summary 235
Job Progress Invoices vs. Estimates
..................................239-240
Jobs, Time & Mileage projects 234
Labor Cost by Job 224-226
modifying 213-216
Open Purchase Orders by Job ...241-242
payables 175-190
profit and loss statements ..256-258, 259
QuickBooks Premier Contractor
Edition 244-246
Retentions Receivable 222-223

Contractor's Guide to QuickBooks Pro **339**

Index

setting QuickBooks Pro preferences
...................................46-47
Time by Job 241
Time by Job Detail 242-243
Time by Name Report 243-244
Transaction Detail by Account ...249-250
unbilled costs 240-241
Unpaid Job Bills 229-232, 246
using project reports 234-246
using sample memorized reports
..................................217-234
Vendor Comp Expirations
.......................186-187, 233-234
Workers' Compensation 226-227
Retainage 164-168
invoicing 168
setting up retentions account in
Chart of Accounts 164-167
setting up retentions item 164-165
Retainage, retentions, viewing
retainage amounts 167
Retained Earnings, posting a
transaction 35-36
Retention, invoicing 166-168
Retentions receivable 164-168
Retentions Receivable report 222-223
Revenue, Chart of Accounts 55-57

S

Sale of property, recording invoice
..................................283-284
Sales tax
converting estimates to invoices 305
QuickBooks Pro estimate 289-290, 296
setting QuickBooks Pro preferences
...................................48-49
Sales, setting QuickBooks Pro
preferences 48-49
Sample company file 19
sample.qbw 144
Chart of Accounts 57-62, 85
Chart of Accounts report 218-219
construction categories 287, 306
contents 19-20
Item List 76, 288
Job Cost Class report 104, 220-222
job divisions in 287, 306
Labor Cost by Job report 224-226
locating 20-21
memorized reports 217-234
payroll, processing 191
Unpaid Job Bills report 229-232
Vendor Comp Expirations report
..................................233-234
Workers' Comp Expires 186
Workers' Compensation report ...226-227
Selecting a QuickBooks Program 7-8
Service Connection, setting QuickBooks
Pro preferences 50
Service items 76-78
Set price invoices 152-154
Set up (employee) card, sole proprietor
or partner 127-128
Setting preferences for estimating
..................................289-290

Show Me, tutorial video 12, 17, 308
Sole proprietorship
adding owner costs to jobs 201-204
Chart of Accounts 56
set up (employee) card 127-128
setting up payroll items 96-102
timesheets 201-204
Specialty contracting, estimates ...288-291
Spelling, setting QuickBooks Pro
preferences 51
Subclasses 105
Subcontractors 115-119
checking workers' comp expiration
..................................186-187
entering opening balance bills for
Accounts Payable 132-133
estimates 288-299
IRS 1099 forms 115-119
organizing files for 138
setting QuickBooks Pro preferences
...................................43-45
setting up non-1099 subcontractors
..................................119
using custom fields 188-190
Subitems..........................77-78
Subtotals, QuickBooks Pro estimate 294
Subtypes 112-113
Summary estimates 144-146
System requirements of, Contractor's
Guide CD-ROM 13, 309

T

Tax tables, obtaining 12
Taxes, land development 278
Technical support
Craftsman Book Company ...14, 302, 304
Intuit 14
Job Cost Wizard 14, 304
National Estimator 14, 304
QuickBooks Pro 14
Time and materials invoices 155-157
billing for employee labor 194
Time by Job Detail Report 242-243
Time by Job Report 241
Time by Name Report 243-244
Time recording 11-12
Time tracking, setting QuickBooks Pro
preferences 54
Timesheets
correcting 198
entering 191-194
missing payroll items on 198
organizing 139-141
owner/partner 201-204
Tracking
change orders 162-164
committed costs (buyouts) 177-180
job costs through QuickBooks Pro
..................................302-307
multiple payment draws 177-179
owner's time for job costs 127-128
sales tax in estimates 296
setting audit trail tracking preferences
...................................35
setting class tracking preferences 104

setting time tracking preferences 54
unfilled purchase orders 180-181
using classes to track cost categories
..................................103
using payroll items to track workers'
comp costs 85-93
Transaction Detail by Account report
..................................249-250
Transactions, flow chart 137
Trial version, QuickBooks Pro 8-9, 17
Try It/Buy It, Web downloads 301-302
Tutorial, Show Me 12, 17, 308

U

Uninstalling Contractor's Guide
CD-ROM 14, 309
Unpaid Bills, by job 246
Unpaid Job Bills report 229-232, 246
Upgrading
QuickBooks older version 23-25
Quicken data files 25-30

V

Vendor Comp Expirations report
.......................186-187, 233-234
Vendor report, preparing on custom
field 188-190
Vendors 115-119
checking workers' comp expiration
..................................186-187
entering opening balance bills for
Accounts Payable 132-133
IRS 1099 forms 115-119
organizing files for 138
setting QuickBooks Pro preferences
...................................43-45
setting up non-1099 vendors 119
updating 28
using custom fields 188-190

W

Web downloads, database and
estimating software 301-302
Web sites
Craftsman Book Company 286, 302
Intuit 8, 17, 286
Online Accounting 9, 286
Work flow, organizing 135-141
Work in Process account 262
Workers' Compensation
adding employees workers' comp
cost 199-200
checking expiration 186-187, 233-234
payroll items for company
contribution 85-93
payroll items for workers' comp
classifications 86-89
payroll items you don't track 93-95
tracking costs 85-93
Vendor Comp Expirations
report 233-234
Workers' Compensation report 226-227

Practical References for Builders

Basic Plumbing with Illustrations, Revised

This completely-revised edition brings this comprehensive manual fully up-to-date with all the latest plumbing codes. It is the journeyman's and apprentice's guide to installing plumbing, piping, and fixtures in residential and light commercial buildings: how to select the right materials, lay out the job and do professional-quality plumbing work, use essential tools and materials, make repairs, maintain plumbing systems, install fixtures, and add to existing systems. Includes extensive study questions at the end of each chapter, and a section with all the correct answers. **384 pages, 8½ x 11, $33.00**

CD Estimator

If your computer has *Windows*™ and a CD-ROM drive, *CD Estimator* puts at your fingertips 85,000 construction costs for new construction, remodeling, renovation & insurance repair, electrical, plumbing, HVAC and painting. Quarterly cost updates are available at no charge on the Internet. You'll also have the *National Estimator* program — a stand-alone estimating program for *Windows*™ that *Remodeling* magazine called a "computer wiz," and Job Cost Wizard, a program that lets you export your estimates to *QuickBooks Pro* for actual job costing. A 60-minute interactive video teaches you how to use this CD-ROM to estimate construction costs. And to top it off, to help you create professional-looking estimates, the disk includes over 40 construction estimating and bidding forms in a format that's perfect for nearly any *Windows*™ word processing or spreadsheet program.
CD Estimator is $73.50

Profits in Buying & Renovating Homes

Step-by-step instructions for selecting, repairing, improving, and selling highly profitable "fixer-uppers." Shows which price ranges offer the highest profit-to-investment ratios, which neighborhoods offer the best return, practical directions for repairs, and tips on dealing with buyers, sellers, and real estate agents. Shows you how to determine your profit before you buy, what "bargains" to avoid, and how to make simple, profitable, inexpensive upgrades. **304 pages, 8½ x 11, $24.75**

Basic Concrete Engineering for Builders

Basic concrete design principles in terms readily understood by anyone who has poured and finished site-cast structural concrete. Shows how structural engineers design concrete for buildings – foundations, slabs, columns, walls, girders, and more. Tells you what you need to know about admixtures, reinforcing, and methods of strengthening concrete, plus tips on field mixing, transit mix, pumping, and curing. Explains how to design forms for maximum strength and to prevent blow-outs, form and size slabs, beams, columns and girders, calculate the right size and reinforcing for foundations, figure loads and carrying capacities, design concrete walls, and more. Includes a CD-ROM with a limited version of an engineering software program to help you calculate beam, slab and column size and reinforcement. **256 pages, 8½ x 11, $39.50**

National Painting Cost Estimator

A complete guide to estimating painting costs for just about any type of residential, commercial, or industrial painting, whether by brush, spray, or roller. Shows typical costs and bid prices for fast, medium, and slow work, including material costs per gallon; square feet covered per gallon; square feet covered per manhour; labor, material, overhead, and taxes per 100 square feet; and how much to add for profit. Includes a CD-ROM with an electronic version of the book with *National Estimator*, a stand-alone *Windows*™ estimating program, plus an interactive multimedia video that shows how to use the disk to compile construction cost estimates.
440 pages, 8½ x 11, $53.00. Revised annually

Blueprint Reading for the Building Trades

How to read and understand construction documents, blueprints, and schedules. Includes layouts of structural, mechanical, HVAC and electrical drawings. Shows how to interpret sectional views, follow diagrams and schematics, and covers common problems with construction specifications.
192 pages, 5½ x 8½, $14.75

2005 *National Electrical Code*

This new electrical code incorporates sweeping improvements to make the code more functional and user-friendly. Here you'll find the essential foundation for electrical code requirements for the 21st century. With hundreds of significant and widespread changes, this 2005 *NEC* contains all the latest electrical technologies, recently developed techniques, and enhanced safety standards for electrical work. This is the standard all electricians are required to know, even if it hasn't yet been adopted by their local or state jurisdictions. **784 pages, 8½ x 11, $65.00**

National Construction Estimator

Current building costs for residential, commercial, and industrial construction. Estimated prices for every common building material. Provides manhours, recommended crew, and gives the labor cost for installation. Includes a CD-ROM with an electronic version of the book with *National Estimator*, a stand-alone *Windows*™ estimating program, plus an interactive multimedia video that shows how to use the disk to compile construction cost estimates. **656 pages, 8½ x 11, $52.50. Revised annually**

Contractor's Guide to the Building Code Revised

This new edition was written in collaboration with the International Conference of Building Officials, writers of the code. It explains in plain English exactly what the latest edition of the *Uniform Building Code* requires. Based on the 1997 code, it explains the changes and what they mean for the builder. Also covers the *Uniform Mechanical Code* and the *Uniform Plumbing Code*. Shows how to design and construct residential and light commercial buildings that'll pass inspection the first time. Suggests how to work with an inspector to minimize construction costs, what common building shortcuts are likely to be cited, and where exceptions may be granted. **320 pages, 8½ x 11, $39.00**

Construction Forms & Contracts

125 forms you can copy and use — or load into your computer (from the FREE disk enclosed). Then you can customize the forms to fit your company, fill them out, and print. Loads into *Word* for *Windows*™, *Lotus 1-2-3*, *WordPerfect*, *Works*, or *Excel* programs. You'll find forms covering accounting, estimating, fieldwork, contracts, and general office. Each form comes with complete instructions on when to use it and how to fill it out. These forms were designed, tested and used by contractors, and will help keep your business organized, profitable and out of legal, accounting and collection troubles. Includes a CD-ROM for *Windows*™ and Mac.
432 pages, 8½ x 11, $41.75

Basic Engineering for Builders

If you've ever been stumped by an engineering problem on the job, yet wanted to avoid the expense of hiring a qualified engineer, you should have this book. Here you'll find engineering principles explained in non-technical language and practical methods for applying them on the job. With the help of this book you'll be able to understand engineering functions in the plans and how to meet the requirements, how to get permits issued without the help of an engineer, and anticipate requirements for concrete, steel, wood and masonry. See why you sometimes have to hire an engineer and what you can undertake yourself: surveying, concrete, lumber loads and stresses, steel, masonry, plumbing, and HVAC systems. This book is designed to help the builder save money by understanding engineering principles that you can incorporate into the jobs you bid. **400 pages, 8½ x 11, $36.50**

Contractor's Survival Manual

How to survive hard times and succeed during the up cycles. Shows what to do when the bills can't be paid, finding money and buying time, transferring debt, and all the alternatives to bankruptcy. Explains how to build profits, avoid problems in zoning and permits, taxes, time-keeping, and payroll. Unconventional advice on how to invest in inflation, get high appraisals, trade and postpone income, and stay hip-deep in profitable work. **160 pages, 8½ x 11, $25.00**

Contracting in All 50 States

Every state has its own licensing requirements that you must meet to do business there. These are usually written exams, financial requirements, and letters of reference. This book shows how to get a building, mechanical or specialty contractor's license, qualify for DOT work, and register as an out-of-state corporation, for every state in the U.S. It lists addresses, phone numbers, application fees, requirements, where an exam is required, what's covered on the exam and how much weight each area of construction is given on the exam. You'll find just about everything you need to know in order to apply for your out-of-state license. **416 pages, 8½ x 11, $36.00**

How to Succeed With Your Own Construction Business

Everything you need to start your own construction business: setting up the paperwork, finding the work, advertising, using contracts, dealing with lenders, estimating, scheduling, finding and keeping good employees, keeping the books, and coping with success. If you're considering starting your own construction business, all the knowledge, tips, and blank forms you need are here. **336 pages, 8½ x 11, $28.50**

Basic Lumber Engineering for Builders

Beam and lumber requirements for many jobs aren't always clear, especially with changing building codes and lumber products. Most of the time you rely on your own "rules of thumb" when figuring spans or lumber engineering. This book can help you fill the gap between what you can find in the building code span tables and what you need to pay a certified engineer to do. With its large, clear illustrations and examples, this book shows you how to figure stresses for pre-engineered wood or wood structural members, how to calculate loads, and how to design your own girders, joists and beams. Included FREE with the book — an easy-to-use limited version of NorthBridge Software's *Wood Beam Sizing* program.
272 pages, 8½ x 11, $38.00

National Renovation & Insurance Repair Estimator

Current prices in dollars and cents for hard-to-find items needed on most insurance, repair, remodeling, and renovation jobs. All price items include labor, material, and equipment breakouts, plus special charts that tell you exactly how these costs are calculated. Includes a CD-ROM with an electronic version of the book with *National Estimator*, a stand-alone Windows™ estimating program, plus an interactive multimedia video that shows how to use the disk to compile construction cost estimates.
568 pages, 8½ x 11, $54.50. Revised annually

Estimating Home Building Costs

Estimate every phase of residential construction from site costs to the profit margin you include in your bid. Shows how to keep track of manhours and make accurate labor cost estimates for footings, foundations, framing and sheathing finishes, electrical, plumbing, and more. Provides and explains sample cost estimate worksheets with complete instructions for each job phase. **320 pages, 5½ x 8½, $17.00**

Plumber's Handbook Revised

This new edition shows what will and won't pass inspection in drainage, vent, and waste piping, septic tanks, water supply, graywater recycling systems, pools and spas, fire protection, and gas piping systems. All tables, standards, and specifications are completely up-to-date with recent plumbing code changes. Covers common layouts for residential work, how to size piping, select and hang fixtures, practical recommendations, and trade tips. It's the approved reference for the plumbing contractor's exam in many states. Includes an extensive set of multiple choice questions after each chapter, and in the back of the book, the answers and explanations. Also in the back of the book, a full sample plumber's exam.
352 pages, 8½ x 11, $32.00

Roof Framing

Shows how to frame any type of roof in common use today, even if you've never framed a roof before. Includes using a pocket calculator to figure any common, hip, valley, or jack rafter length in seconds. Over 400 illustrations cover every measurement and every cut on each type of roof: gable, hip, Dutch, Tudor, gambrel, shed, gazebo, and more.
480 pages, 5½ x 8½, $24.50

Troubleshooting Guide to Residential Construction

How to solve practically every construction problem – before it happens to you! With this book you'll learn from the mistakes other builders made as they faced 63 typical residential construction problems. Filled with clear photos and drawings that explain how to enhance your reputation as well as your bottom line by avoiding problems that plague most builders. Shows how to avoid, or fix, problems ranging from defective slabs, walls and ceilings, through roofing, plumbing & HVAC, to paint.
304 pages, 8½ x 11, $32.50

National Electrical Estimator

This year's prices for installation of all common electrical work: conduit, wire, boxes, fixtures, switches, outlets, loadcenters, panelboards, raceway, duct, signal systems, and more. Provides material costs, manhours per unit, and total installed cost. Explains what you should know to estimate each part of an electrical system. Includes a CD-ROM with an electronic version of the book with *National Estimator*, a stand-alone Windows™ estimating program, plus an interactive multimedia video that shows how to use the disk to compile construction cost estimates.
552 pages, 8½ x 11, $52.75. Revised annually

Construction Estimating Reference Data

Provides the 300 most useful manhour tables for practically every item of construction. Labor requirements are listed for sitework, concrete work, masonry, steel, carpentry, thermal and moisture protection, doors and windows, finishes, mechanical and electrical. Each section details the work being estimated and gives appropriate crew size and equipment needed. Includes a CD-ROM with an electronic version of the book with *National Estimator*, a stand-alone Windows™ estimating program, plus an interactive multimedia video that shows how to use the disk to compile construction cost estimates. **384 pages, 11 x 8½, $39.50**

Build Smarter with Alternative Materials

New building products are coming out almost every week. Some of them may become new standards, as sheetrock replaced lath and plaster some years ago. Others are little more than a gimmick. To write this manual, the author researched hundreds of products that have come on the market in recent years. The ones he describes in this book will do the job better, creating a superior, longer-lasting finished product, and in many cases also save you time and money. Some are made with recycled products— a good selling point with many customers. But most of all, they give you choices, so you can give your customers choices. In this book, you'll find materials for almost all areas of constructing a house, from the ground up. For each product described, you'll learn where you can get it, where to use it, what benefits it provides, any disadvantages, and how to install it — including tips from the author. And to help you price your jobs, each description ends with manhours – for both the first time you install it, and after you've done it a few times. **336 pages, 8½ x 11, $34.75**

Moving to Commercial Construction

In commercial work, a single job can keep you and your crews busy for a year or more. The profit percentages are higher, but so is the risk involved. This book takes you step-by-step through the process of setting up a successful commercial business; finding work, estimating and bidding, value engineering, getting through the submittal and shop drawing process, keeping a stable work force, controlling costs, and promoting your business. Explains the design/build and partnering business concepts and their advantage over the competitive bid process. Includes sample letters, contracts, checklists and forms that you can use in your business, plus a CD-ROM with blank copies in several word-processing formats for both Mac and PC computers. **256 pages, 8½ x 11, $42.00**

Estimating Excavation

How to calculate the amount of dirt you'll have to move and the cost of owning and operating the machines you'll do it with. Detailed, step-by-step instructions on how to assign bid prices to each part of the job, including labor and equipment costs. Also, the best ways to set up an organized and logical estimating system, take off from contour maps, estimate quantities in irregular areas, and figure your overhead.
448 pages, 8½ x 11, $39.50

Building Contractor's Exam Preparation Guide

Passing today's contractor's exams can be a major task. This book shows you how to study, how questions are likely to be worded, and the kinds of choices usually given for answers. Includes sample questions from actual state, county, and city examinations, plus a sample exam to practice on. This book isn't a substitute for the study material that your testing board recommends, but it will help prepare you for the types of questions — and their correct answers — that are likely to appear on the actual exam. Knowing how to answer these questions, as well as what to expect from the exam, can greatly increase your chances of passing.
320 pages, 8½ x 11, $35.00

National Home Improvement Estimator

Current labor and material prices for home improvement projects. Provides manhours for each job, recommended crew size, and the labor cost for the removal and installation work. Material prices are current, with location adjustment factors and free monthly updates on the Web. Gives step-by-step instructions for the work, with helpful diagrams, and home improvement shortcuts and tips from an expert. Includes a CD-ROM with an electronic version of the book, and *National Estimator*, a stand-alone *Windows*™ estimating program, plus an interactive multimedia tutorial that shows how to use the disk to compile home improvement cost estimates. **504 pages, 8½ x 11, $53.75**

National Repair & Remodeling Estimator

The complete pricing guide for dwelling reconstruction costs. Reliable, specific data you can apply on every repair and remodeling job. Up-to-date material costs and labor figures based on thousands of jobs across the country. Provides recommended crew sizes; average production rates; exact material, equipment, and labor costs; a total unit cost and a total price including overhead and profit. Separate listings for high- and low-volume builders, so prices shown are specific for any size business. Estimating tips specific to repair and remodeling work to make your bids complete, realistic, and profitable. Includes a CD-ROM with an electronic version of the book with *National Estimator*, a stand-alone *Windows*™ estimating program, plus an interactive multimedia video that shows how to use the disk to compile construction cost estimates. **296 pages, 8½ x 11, $53.50. Revised annually**

Contractor's Plain-English Legal Guide

For today's contractors, legal problems are like snakes in the swamp – you might not see them, but you know they're there. This book tells you where the snakes are hiding and directs you to the safe path. With the directions in this easy-to-read handbook you're less likely to need a $200-an-hour lawyer. Includes simple directions for starting your business, writing contracts that cover just about any eventuality, collecting what's owed you, filing liens, protecting yourself from unethical subcontractors, and more. For about the price of 15 minutes in a lawyer's office, you'll have a guide that will make many of those visits unnecessary. Includes a CD-ROM with blank copies of all the forms and contracts in the book.
272 pages, 8½ x 11, $49.50

Getting Financing & Developing Land

Developing land is a major leap for most builders – yet that's where the big money is made. This book gives you the practical knowledge you need to make that leap. Learn how to prepare a market study, select a building site, obtain financing, guide your plans through approval, then control your building costs so you can ensure yourself a good profit. Includes a CD-ROM with forms, checklists, and a sample business plan you can customize and use to help you sell your idea to lenders and investors.
232 pages, 8½ x 11, $39.00

Craftsman's Construction Installation Encyclopedia

Step-by-step installation instructions for just about any residential construction, remodeling or repair task, arranged alphabetically, from Acoustic tile to Wood flooring. Includes hundreds of illustrations that show how to build, install, or remodel each part of the job, as well as manhour tables for each work item so you can estimate and bid with confidence. Also includes a CD-ROM with all the material in the book, handy look-up features, and the ability to capture and print out for your crew the instructions and diagrams for any job. **792 pages, 8½ x 11, $65.00**

Home Builder Contract and Management Forms

93 essential contracts and forms from the NAHB to help you systemize your business, build more productively, and save hours of work. Here you'll find forms for sales and marketing, contracts and agreements, trade contractor agreements and checklists, construction management forms, and walk through and warranty forms for maintaining your final quality control checks and warranty work. Includes a CD-ROM with all the forms you can use to customize for your company. **316 pages, 8½ x 11, $62.50**

Paint Contractor's Manual

How to start and run a profitable paint contracting company: getting set up and organized to handle volume work, avoiding mistakes, squeezing top production from your crews and the most value from your advertising dollar. Shows how to estimate all prep and painting. Loaded with manhour estimates, sample forms, contracts, charts, tables and examples you can use. **224 pages, 8½ x 11, $28.50**

Steel-Frame House Construction

Framing with steel has obvious advantages over wood, yet building with steel requires new skills that can present challenges to the wood builder. This new book explains the secrets of steel framing techniques for building homes, whether pre-engineered or built stick by stick. It shows you the techniques, the tools, the materials, and how you can make it happen. Includes hundreds of photos and illustrations, plus a CD-ROM with steel framing details, a database of steel materials and manhours, with an estimating program. **320 pages, 8½ x 11, $39.75**

Contractor's Index to the 1997 *Uniform Building Code*

Finally, there's a common-sense index that helps you quickly and easily find the section you're looking for in the *UBC*. It lists topics under the names builders actually use in construction. Best of all, it gives the full section number and the actual page in the *UBC* where you'll find it. If you need to know the requirements for windows in exit access corridor walls, just look under Windows. You'll find the requirements you need are in Section 1004.3.4.3.2.2 in the *UBC* — on page 1-115. This practical index was written by a former builder and building inspector who knows the *UBC* from both perspectives. If you hate to spend valuable time hunting through pages of fine print for the information you need, this is the book for you.
192 pages, 8½ x 11, Loose-leaf, $29.00

Estimating Electrical Construction

Like taking a class in how to estimate materials and labor for residential and commercial electrical construction. Written by an A.S.P.E. National Estimator of the Year, it teaches you how to use labor units, the plan take-off, and the bid summary to make an accurate estimate, how to deal with suppliers, use pricing sheets, and modify labor units. Provides extensive labor unit tables and blank forms for your next electrical job.
272 pages, 8½ x 11, $35.00

The Contractor's Legal Kit

Stop "eating" the costs of bad designs, hidden conditions, and job surprises. Set ground rules that assign those costs to the rightful party ahead of time. And it's all in plain English, not "legalese." For less than the cost of an hour with a lawyer you'll learn the exclusions to put in your agreements, why your insurance company may pay for your legal defense, how to avoid liability for injuries to your sub and his employees or damages they cause, how to collect on lawsuits you win, and much more. It also includes a FREE computer disk with contracts and forms you can customize for your own use. **352 pages, 8½ x 11, $69.95**

Craftsman's Illustrated Dictionary of Construction Terms

Almost everything you could possibly want to know about any word or technique in construction. Hundreds of up-to-date construction terms, materials, drawings and pictures with detailed, illustrated articles describing equipment and methods. Terms and techniques are explained or illustrated in vivid detail. Use this valuable reference to check spelling, find clear, concise definitions of construction terms used on plans and construction documents, or learn about little-known tools, equipment, tests and methods used in the building industry. It's all here.
416 pages, 8½ x 11, $36.00

Home Inspection Handbook

Every area you need to check in a home inspection — especially in older homes. Twenty complete inspection checklists: building site, foundation and basement, structural, bathrooms, chimneys and flues, ceilings, interior & exterior finishes, electrical, plumbing, HVAC, insects, vermin and decay, and more. Also includes information on starting and running your own home inspection business. **324 pages, 5½ x 8½, $24.95**

Estimating With Microsoft *Excel*

Most builders estimate with *Excel* because it's easy to learn, quick to use, and can be customized to your style of estimating. Here you'll find step-by-step instructions on how to create your own customized automated spreadsheet estimating program for use with *Excel*. You'll learn how to use the magic of *Excel* to create detail sheets, cost breakdown summaries, and links. You'll put this all to use in estimating concrete, rebar, permit fees, and roofing. You can even create your own macros. Includes a CD-ROM that illustrates examples in the book and provides you with templates you can use to set up your own estimating system. **148 pages, 7 x 9, $39.95**

National Building Cost Manual

Square foot costs for residential, commercial, industrial, and farm buildings. Quickly work up a reliable budget estimate based on actual materials and design features, area, shape, wall height, number of floors, and support requirements. Includes all the important variables that can make any building unique from a cost standpoint.
240 pages, 8½ x 11, $28.00. Revised annually

Contractor's Year-Round Tax Guide Revised

How to set up and run your construction business to minimize taxes: corporate tax strategy and how to use it to your advantage, and what you should be aware of in contracts with others. Covers tax shelters for builders, write-offs and investments that will reduce your taxes, accounting methods that are best for contractors, and what the I.R.S. allows and what it often questions. **208 pages, 8½ x 11, $26.50**

Builder's Guide to Accounting Revised

Step-by-step, easy-to-follow guidelines for setting up and maintaining records for your building business. This practical guide to all accounting methods shows how to meet state and federal accounting requirements, explains the new depreciation rules, and describes how the Tax Reform Act can affect the way you keep records. Full of charts, diagrams, simple directions and examples, to help you keep track of where your money is going. Recommended reading for many state contractor's exams. Each chapter ends with a set of test questions, and a CD-ROM included FREE has all the questions in interactive self-test software. Use the Study Mode to make studying for the exam much easier, and Exam Mode to practice your skills. **356 pages, 8½ x 11, $35.50**

Rough Framing Carpentry

If you'd like to make good money working outdoors as a framer, this is the book for you. Here you'll find shortcuts to laying out studs; speed cutting blocks, trimmers and plates by eye; quickly building and blocking rake walls; installing ceiling backing, ceiling joists, and truss joists; cutting and assembling hip trusses and California fills; arches and drop ceilings — all with production line procedures that save you time and help you make more money. Over 100 on-the-job photos of how to do it right and what can go wrong. **304 pages, 8½ x 11, $26.50**

Markup & Profit: A Contractor's Guide

In order to succeed in a construction business, you have to be able to price your jobs to cover all labor, material and overhead expenses, and make a decent profit. The problem is knowing what markup to use. You don't want to lose jobs because you charge too much, and you don't want to work for free because you've charged too little. If you know how to calculate markup, you can apply it to your job costs to find the right sales price for your work. This book gives you tried and tested formulas, with step-by-step instructions and easy-to-follow examples, so you can easily figure the markup that's right for your business. Includes a CD-ROM with forms and checklists for your use. **320 pages, 8½ x 11, $32.50**

Craftsman Book Company
6058 Corte del Cedro
P.O. Box 6500
Carlsbad, CA 92018

☎ 24 hour order line
1-800-829-8123
Fax (760) 438-0398

In A Hurry?
We accept phone orders charged to your
○ Visa, ○ MasterCard, ○ Discover or ○ American Express

Card#_____

Exp. date_____ Initials_____

Tax Deductible: Treasury regulations make these references tax deductible when used in your work. Save the canceled check or charge card statement as your receipt.

Order online http://www.craftsman-book.com
Free on the Internet! Download any of Craftsman's estimating costbooks for a 30-day free trial! http://costbook.com

Name _____
e-mail address (for order tracking and special offers)
Company _____
Address _____
City/State/Zip _____ ○ This is a residence
Total enclosed_____ (In California add 7.25% tax)
We pay shipping when your check covers your order in full.

10-Day Money Back Guarantee

- ○ 39.50 Basic Concrete Engineering for Builders
- ○ 36.50 Basic Engineering for Builders
- ○ 38.00 Basic Lumber Engineering for Builders
- ○ 33.00 Basic Plumbing with Illustrations Revised
- ○ 14.75 Blueprint Reading for the Building Trades
- ○ 34.75 Build Smarter with Alternative Materials
- ○ 35.50 Builder's Guide to Accounting Revised
- ○ 35.00 Building Contractor's Exam Prep Guide
- ○ 73.50 CD Estimator
- ○ 39.50 Construction Estimating Reference Data with FREE National Estimator on a CD-ROM.
- ○ 41.75 Construction Forms & Contracts with a CD-ROM for Windows™ and Macintosh.
- ○ 36.00 Contracting in All 50 States
- ○ 39.00 Contractor's Guide to Building Code Rev
- ○ 29.00 Contractor's Index to the UBC - Looseleaf
- ○ 69.95 Contractor's Legal Kit
- ○ 49.50 Contractor's Plain-English Legal Guide
- ○ 25.00 Contractor's Survival Manual
- ○ 26.50 Contractor's Year-Round Tax Guide Revised

- ○ 65.00 Craftsman's Construction Installation Encyclopedia
- ○ 36.00 Craftsman's Illustrated Dictionary of Construction Terms
- ○ 35.00 Estimating Electrical Construction
- ○ 39.50 Estimating Excavation
- ○ 17.00 Estimating Home Building Costs
- ○ 39.95 Estimating with Microsoft Excel
- ○ 39.00 Getting Financing & Developing Land
- ○ 62.50 Home Builder Contract and Management Forms
- ○ 24.95 Home Inspection Handbook
- ○ 28.50 How to Succeed w/Your Own Construction Business
- ○ 32.50 Markup & Profit: A Contractor's Guide
- ○ 42.00 Moving to Commercial Construction
- ○ 28.00 National Building Cost Manual
- ○ 52.50 National Construction Estimator with FREE National Estimator on a CD-ROM.
- ○ 65.00 2005 National Electrical Code
- ○ 52.75 National Electrical Estimator with FREE National Estimator on a CD-ROM.

- ○ 53.75 National Home Improvement Estimator w/ FREE National Estimator CD.
- ○ 53.00 National Painting Cost Estimator with FREE National Estimator on a CD-ROM.
- ○ 54.50 National Renovation & Insurance Repair Estimator w/ FREE National Estimator CD.
- ○ 53.50 National Repair & Remodeling Estimator with FREE National Estimator on a CD-ROM.
- ○ 28.50 Paint Contractor's Manual
- ○ 32.00 Plumber's Handbook Revised
- ○ 24.75 Profits in Buying & Renovating Homes
- ○ 24.50 Roof Framing
- ○ 26.50 Rough Framing Carpentry
- ○ 39.75 Steel-Frame House Construction
- ○ 32.50 Troubleshooting Guide to Res Construct
- ○ 45.25 Contractor's Guide to QuickBooks Pro 2001
- ○ 47.75 Contractor's Guide to QuickBooks Pro 2003
- ○ 48.50 Contractor's Guide to QuickBooks Pro 2004
- ○ 49.75 Contractor's Guide to QuickBooks Pro 2005
- ○ FREE Full Color Catalog

Prices subject to change without notice

QuickBooks Help!

Online ACCOUNTING CAN PROVIDE:
- ✔ Online Support
- ✔ Telephone Support
- ✔ Onsite Support
- ✔ Seminars

Call **888-254-9252** or visit us on the web at:
www.onlineaccounting.com

☐ Please e-mail more information to me at:
_____

☐ Please fax more information to me at:
(_____) _____

☐ Please send more information to me at:
_____
NAME
_____
COMPANY
_____
STREET ADDRESS
_____
CITY, STATE, ZIP

GC/WORKS — SOFTWARE FOR THE SMART BUILDER

Designed specifically for the small to mid size construction company

- Homebuilder Edition
- Standard Edition
- Contractor Edition
- Developer Edition
- Lite Edition

WINDOWS & MACINTOSH

www.synapsesoftware.com

SYNAPSE SOFTWARE INC.

* Estimating
* Job Costing
* Accounting
* Scheduling
* Purchase Orders
* Change Orders
* Subcontracts
* Options & Extras
* Payroll

Powerful Simple & More

Uses QuickBooks Pro & AppleWorks

30 Day Money Back Guarantee

CALL TODAY 800-420-2521

YES -- I want to find out more about the hottest new software tools for the construction industry. Please rush me all of the details on GC/Works.

Name _____
Company _____
Street _____
City-St-Zip _____
Phone _____ Fax _____

Mail This Card Today
For a Free Full Color Catalog

Over 100 books, annual cost guides and estimating software packages at your fingertips with information that can save you time and money. Here you'll find information on carpentry, contracting, estimating, remodeling, electrical work, and plumbing.

All items come with an unconditional 10-day money-back guarantee. If they don't save you money, mail them back for a full refund.

Name _____
Company _____
Address _____
City/State/Zip _____

Craftsman Book Company / 6058 Corte del Cedro / P.O. Box 6500 / Carlsbad, CA 92018

PLACE STAMP HERE

ONLINE ACCOUNTING
P.O. Box 873
Eagle, ID 83616

Place Stamp Here

PO Box 845
Henrietta, NY 14467

BUSINESS REPLY MAIL
FIRST CLASS MAIL PERMIT NO. 271 CARLSBAD, CA

POSTAGE WILL BE PAID BY ADDRESSEE

NO POSTAGE NECESSARY IF MAILED IN THE UNITED STATES

 Craftsman Book Company
6058 Corte del Cedro
P.O. Box 6500
Carlsbad, CA 92018-9974